T0235653

Transdisciplinary Thinking from the Global South

This book promotes constructive and nuanced transdisciplinary understandings of some of the critical problems that we face on a global scale today by thinking with and from the Global South. It is engaged in transmodernising, pluriversalising, decolonising, queering, and/or posthumanising thinking and practice.

The book aims to contribute to and challenge current debates regarding knowledge, diversity, and change. This is achieved through the application of transdisciplinary and indisciplined perspectives to the Himalayan Anthropocene; transport services in Mexico City; the EU-Turkey border regimes and policy; egoism and the decolonisation of whiteness; the Witch and the decolonisation of the gender binary; Nepalese students in Denmark; and the decolonisation of global health promotion. The book thereby provides the reader a multiplicity of pathways of knowledges and practices that address current problems co-produced by the dominant Western colonial onto-epistemic outset, giving way to 'other' knowledge-practices, towards a pluriversal approach.

This book will be of interest to upper-level undergraduate and graduate students in disciplines such as human geography, development studies, politics, international relations, sociology, anthropology, cultural studies, planning, and philosophy. It is also relevant to researchers, development workers and human rights/environmental activists, and other intellectual practitioners.

Juan Carlos Finck Carrales is a Lecturer of Social Sciences and Urban Planning at Roskilde University (RUC) in Denmark. He teaches and supervises research at the International Bachelor in Social Sciences, the Bachelor in Global Humanities and the Master of Nordic Urban Planning Studies (NUPS). He also coordinates the Language Profile Program of RUC where he is responsible for the Spanish Language Profile. He is part of the Mobility, Space, Place, and Urban Studies (MOSPUS) research group. He holds a PhD degree in Social Sciences from the Program of Society, Space and Technology of RUC. His inter- and transdisciplinary research intersects the fields of Policymaking, Mobility, Decolonial Studies, and Urban and Transport Planning by making the use of ethnographic, participatory, interactive and mixed methods. He has participated in urban projects and formalization processes of transport services in Mexico City, whose outcomes have been reproduced in the media and have influenced the policymaking and regulation of the city. He is a consultant for the Integral Mobility Program of Mexico City 2020-2024.

Julia Suárez-Krabbe is Associate Professor in Cultural Encounters at the Department of Communication and Arts, Roskilde University, Denmark, and Distinguished Research Associate at the Ali Mazrui Centre for Higher Education, University of Johannesburg, South Africa. Her work centers on racism, human rights, development, knowledge production, education and decolonisation in Europe and the Americas. Her latest work includes the co-authorship of the report "Stop Killing Us Slowly. A Research Report on the Motivation Enhancement Measures and the Criminalization of Rejected Asylum Seekers in Denmark" from 2018, which includes examinations of state-sanctioned racism in Danish deportation camps, and was written in collaboration with the refugee movement in Denmark. Her work additionally revolves around the ontological, epistemological and existential dimensions of decolonisation. Julia is the author of "Race, Rights and Rebels. Alternatives to Human Rights and Development from the Global South" (2016).

Routledge Research on Decoloniality and New Postcolonialisms

Series Editor: Mark Jackson, Senior Lecturer in Postcolonial Geographies, School of Geographical Sciences, University of Bristol, UK.

Routledge Research on Decoloniality and New Postcolonialisms is a forum for original, critical research into the histories, legacies, and life-worlds of modern colonialism, postcolonialism, and contemporary coloniality. It analyses efforts to decolonise dominant and damaging forms of thinking and practice, and identifies, from around the world, diverse perspectives that encourage living and flourishing differently. Once the purview of a postcolonial studies informed by the cultural turn's important focus on identity, language, text, and representation, today's resurgent critiques of coloniality are also increasingly informed, across the humanities and social sciences, by a host of new influences and continuing insights for different futures: indigeneity, critical race theory, relational ecologies, critical semiotics, posthumanisms, ontology, affect, feminist standpoints, creative methodologies, post-development, critical pedagogies, intercultural activisms, place-based knowledges, and much else. The series welcomes a range of contributions from socially engaged intellectuals, theoretical scholars, empirical analysts, and critical practitioners whose work attends, and commits, to newly rigorous analyses of alternative proposals for understanding life and living well on our increasingly damaged earth.

This series is aimed at upper-level undergraduates, research students, and academics, appealing to scholars from a range of academic fields including human geography, sociology, politics and broader interdisciplinary fields of social sciences, arts and humanities.

A Decolonial Black Feminist Theory of Reading and Shade
Feeling the University
Andrea N. Baldwin

Transdisciplinary Thinking from the Global South
Whose problems, whose solutions?
Edited by Juan Carlos Finck Carrales and Julia Suárez-Krabbe

For more information about this series, please visit: https://www.routledge.com/Routledge-Research-on-Decoloniality-and-New-Postcolonialisms/book-series/RRNP

Transdisciplinary Thinking from the Global South

Whose Problems, Whose Solutions?

Edited by Juan Carlos Finck Carrales and Julia Suárez-Krabbe

LONDON AND NEW YORK

First published 2022
by Routledge
2 Park Square, Milton Park, Abingdon, Oxon OX14 4RN

and by Routledge
605 Third Avenue, New York, NY 10158

Routledge is an imprint of the Taylor & Francis Group, an informa business

British Library Cataloguing-in-Publication Data
A catalogue record for this book is available from the British Library

Library of Congress Cataloging-in-Publication Data
A catalog record has been requested for this book

ISBN: 978-1-03-200035-0 (hbk)
ISBN: 978-1-03-200038-1 (pbk)
ISBN: 978-1-00-317241-3 (ebk)

DOI: 10.4324/9781003172413

Typeset in Bembo
by KnowledgeWorks Global Ltd.

Contents

List of contributors

Sergejs Asilgarajevs holds a MA in Cultural Encounters and International Development (Roskilde University, Denmark) and a BA in Cultural Encounters and International Studies (Roskilde University, Denmark) with a master thesis on 'De-racialisation of Race in Border Politics; A Decolonial Study of Borders, Race & Knowledge'. With an interest in conflict and security related issues, his ambition is to further research global border politics from a decolonial perspective.

Stephen Carney is Professor of Educational Studies at Roskilde University in Denmark where he leads its Global Humanities study programme. His research focuses on global educational reform and comparative method. He has studied university governance in Denmark, teacher preparation in England and China and school development in Nepal and India. He has been President of the Comparative Education Society in Europe (CESE) since 2016.

Johanne Andersen Elbek holds a Master of Science (MSc) in Health Promotion and Strategies and International Development Studies (Roskilde University, Denmark) and BA in Global Nutrition and Health (VIA University College, Aarhus). She is a research assistant at VIA Research Center for Health and Welfare Technology (Aarhus, Denmark). She has previously researched and published about meal-related challenges in patients with chronic obstructive pulmonary disease. She has a particular interest in working with policies and programmes to improve and promote health outcomes in marginalised groups, both in the Global South and North.

Lene Maj Hjortsø Fernando holds a Master of Science (MSc) in Health Promotion and Strategies & International Development Studies (Roskilde University, Denmark) and BA in Global Nutrition and Health (Metropolitan University College Copenhagen). With emphasis on health being a human right, her ambition is to contribute to creating awareness and conditions that support a healthy and sustainable way of living, in order to improve the quality of humans' prolonged lives. She advocates this agenda must be embedded at all stages of global development; in policies, business strategies and health promotion interventions, not least because investment in health evidently increases socioeconomic development.

Nazila Ghavami Kivi is an Iranian-Danish independent scholar, literary critic, editor and translator, and works in the intersections of academia, activism and art. She holds an MA in Cultural Encounters and Communication (Roskilde University) and a BSc in Public Health Science with Gender Certificate (University of Copenhagen) for her focus on body, gender and minorities. She has worked with Sex Education and Diversity for more than ten years, and had a research year in Social Medicine focusing on reproductive health for minorities in pregnancy and childbirth, and works with the intersections of language, culture and oppression. She has created the course 'From Witches to Cyborgs: Gender, Race and Resistance' at the Danish Institute for Study Abroad (DIS), which ties the history of colonialism, the body and reproduction together.

Avin Mesbah holds a Master of Science (MSc) in International Development Studies and Cultural Encounters of Roskilde University (RUC) and a Bachelor of Arts (BA) in Spanish Language and Culture with electives in Minority Studies and Comparative Cultural Studies of the University of Copenhagen (KU). Her interests are centred around epistemic racism, structural processes of dehumanisation, decolonisation of knowledge, and policymaking—particularly related to refugee rights and mobility. With the master thesis named 'De-racialisation of Race in Border Politics; A Decolonial Study of Borders, Race & Knowledge', she explores the link between the racial injustices at EU border sites and the legacy of colonial knowledge structures, through the example of the EU-Turkey Joint Action Plan of 2015.

Prem Poddar is Professor in Cultural Encounters at Roskilde University in Denmark and Alexander von Humboldt Senior Fellow in Germany. He has been an AHRC Fellow at Southampton and Carlsberg Fellow at Cambridge University. He has taught in India, Britain, and Denmark where he was Associate Professor in Postcolonial Studies. He is the author of many articles and books including 'Violent Civilities' (2002) and 'Postkolonial Contra-modernitet: Immigration, Identitet, Historie' (2004). He has also edited 'Translating Nations' (2000); 'A Historical Companion to Postcolonial Thought' (2005/2008), 'Empire and After: Englishness in Postcolonial Perspective' (2007/2010); 'A Historical Companion to Postcolonial Literatures—Continental Europe and its Empires' (2008/2011) and 'Gorkhas Imagined' (2009). His forthcoming series of articles and a monograph investigate India–China relations through the lens of a border town across the Himalayas. His continuing interest in 'state' and 'nation' as conceptual contexts for analyzing cultural representation forms the centre of his work on the politics of the passport.

Rashmi Singla is an Associate Professor at the Department of People and Technology of Roskilde University in Denmark. She holds a PhD in Psychology Masters (Copenhagen University, Denmark), a Master of Science, and a BSc with Honours (Delhi University, India) in Psychotherapy Specialist. She is also affiliated to NGO—TTT (Transcultural Therapeutic Team for Ethnic Minority Youth and Families). She is currently teaching, researching, participating in international projects and publishing about movements across borders especially on migration, transnationalism, coloniality, Eastern/Western Psychology and family relationships, mental health promotion, and psychosocial intervention. Her last research based book deals with ethnically intermarried couples, mixed parenting and the current one with living apart together transnational (LATT) couples, both related to mental health and well-being promotion.

Nitya Nanda Timsina is an independent scholar based in Denmark. He researches on young people, mobility and higher education within the context of globalisation and international development. He obtained a European Master's degree in Lifelong Learning within the sociology of education and a PhD in Higher Education in Denmark. Prior to coming to Europe, he was a graduate student in Nepal at the Central Department of Political Science and completed secondary school in India. He has taught in secondary schools and been a professional journalist in Kathmandu in addition to performing external consultant tasks for international development organisations in Nepal. He prefers to call himself a thinker, writer and an international vagabond who keeps traveling between Europe and South Asia.

Introduction

Horizons of possibility and scientific research: whose problems, whose solutions?

Juan Carlos Finck Carrales and Julia Suárez-Krabbe

> [...] pueden también corromperse aún pueblos enteros, como cuando la población del Imperio guarda silencio, mira hacia otro lado, ante la inmolación de pueblos inocentes [...].
>
> (Dussel 2006, 47)[1]

> Vi tror ikke på én samlet retning for en anden verden. Men vi tror på, at der er en styrke i, at vi insisterer på, at vores forskellige kampe kun kan vindes, hvis vi møder hinanden og arbejder sammen. For den magt, som truer vores eksistens hænger uløseligt sammen. Vi kan regne med hinanden.
>
> (Marronage 2017, 7)[2]

Read together the quotes above signal some significant changes in terms of imperial/colonial power, resistance, and change. While we still find that many populations in the Global North keep silent and look to the other side before the atrocities itself produces or commits, this book attests to the fact that an increasing number of peoples who find ourselves in the heart of 'Empire' meet each other, stand together, and work for another world. This we do by engaging in the plurality of knowledges, insights and perspectives that emerge when shifting the geography of reason (Gordon 2011), and thinking with and from the Global South. Produced through our engaging in and meeting each other in our diversity of histories, perspectives, problems, and solutions, knowledge can indeed contribute to dismantling the lies of power, whereby the immolation of (our) peoples is justified.

The time difference between the two quotes additionally attest to the fact that the entry of peoples from the Global South into the westernised university around the world has had a significant impact on the social sciences and the humanities. This has produced a tectonic shake, among others by displaying some previously unrecognised core assumptions of the westernised university. One of these pertains to the imbrication between knowledge and colonial power, reflected in the fact that many people still often assume scientific knowledge to be the knowledge produced by white, male, property-owning,

DOI: 10.4324/9781003172413-1

(culturally) Christian elites of a few countries in the world, whose 'science' in turn rests upon the extraction, appropriation, disavowal, or invisibilisation of all other existing knowledges. The question regarding whose problems such idea of science actually addresses, and what problems it disregards in that very same framing is therefore today both a core critical question *to* the dominant social sciences and humanities, and a sort of compass to those of us engaged in enlarging the pathways to another world. While the dominant theories, methodologies, and perspectives require of scholars that we disengage, other-worlding thinking-practices entail developing a relationship to reality and each other that allows us setting new conditions for what is to come (Gordon 2020). Indeed, different communities of knowledge across the globe are increasingly coming together, rejecting the westernised university's monopoly of knowledge, and engaging in debates around pressing issues. These include but are not limited to the decolonisation of the university (Alvares and Faruqi 2012; Bhambra et al. 2018; Boidin, Cohen, and Grosfoguel 2012; Cupples and Grosfoguel 2019; Grosfoguel, Hernández, and Velázquez 2016), climate change and (post-)development (Kothari et al. 2019), international relations and politics (Fiddian-Qasmiyeh and Daley 2019; Frizzo Bragato and Gordon 2017; Rutazibwa and Shilliam 2018; Woons and Weier 2017), and human rights (Barreto 2013; Dhawan 2014).

The increasing number of scholarly work that seeks to decolonise and/or pluriversalise knowledge in different areas attests to the fact that we are living at a historical time of de-linking (Mignolo 2007) from the racist, colonial, capitalist, and patriarchal understanding of knowledge as universal, neutral, objective, and tied to progress. This de-linking is necessarily followed by 'reconstitutions, re-emergence, resurgence, re-existence' (Mignolo and Hoffmann 2017). Such processes involve re-linking in the pluriverse of knowledges and paths tied to specific histories of resistance and co-construction of our worlds beyond the current global crises (Suárez-Krabbe forthcoming).

This book brings together contributions by scholars and independent researchers whose work in different ways is concerned with opening up the canon of knowledge, and forwarding constructive, nuanced understandings of some of the critical problems that we face on a global scale today by thinking with and from the Global South. As such, the question 'whose problems, whose solutions?' is indicative of the central concern present in the contributions compiled here. As implied earlier, such a question is not rhetorical, but significantly an epistemological one. This question can be unpacked in the following way: whose problems does knowledge production address? From where and when, with whom, for whom, for what, and how? (Leyva 2015). Whose problems, then, do the authors whose work we have compiled here address? Like the publications referred to above, this book addresses problems that affect all of humanity, the Earth, and all other-than human beings. It addresses such problems from different epistemological, existential, and socio-historical positions and conditions. All authors in this book live in

Denmark, and have some relationship to Roskilde University. Indeed, the idea for the book emerged out of an elective course that some of the authors offered in the Spring of 2019 and 2020 which aimed to provide students with insights central to current critical knowledge debates (as mentioned above), which are slowly, but powerfully, emerging in academia in Denmark.

The specific Danish context can be characterised as a de facto apartheid-state (Arce and Suárez-Krabbe 2018; Suárez-Krabbe and Lindberg 2019). Here apartheid refers to racism-based policies in any state (Morton 2000), not only to the South African version of the system. In practice, this means, among other things, that different legal and political regimes apply in one and the same state: that is, that there are some laws and rights that apply to some of the citizens, and others that apply to different categories of migrants. Apartheid is seen when some Danes, despite their citizenship, are described as 'non-Western', and on that basis subjected to differential laws and punishment. Apartheid in Denmark is apparent in the so-called 'Ghetto-laws' and the 2018 'Burqa ban', in the 2020 'security package', the 2021 'citizenship rules' and not least in the asylum, refugee, and migration policies that Denmark has pursued for decades (Freedom of Movements Research Collective 2018). Additionally, the public debate is largely characterised by verbal and symbolic attacks on 'non-Western' Danes, especially Muslims (Hassani 2019; Özcan and Bangert 2019). Most of the critique addressed towards these policies, including that coming from academia, does not address the underlying problem, namely state racism. In such a context, Danish academia can appear drastically disengaged, and thereby complicit with the perpetuation and aggravation of these problems. During the spring of 2021 in Denmark, we saw a renewed outbreak of attacks by right-wing politicians and media against critical studies such as those represented in this book. Such attacks are neither new nor country-specific and can take different forms. For instance, in France, such attacks involve the discourse on 'Islamo-leftism' (Nanni and Traverso 2021). In Austria and Greece scholars standing against islamophobia (Siddiqui 2021) and heteropatriarchal normativity (Feminist Autonomous Centre for Research 2021) respectively are under attack, and in the United States the far right is involved in systematic efforts to 'cancel' professors through student-groups (Speri 2021). Accompanied by budget cuts to the social sciences and humanities over the years, in Denmark, such attacks have effectively 'cancelled' many scholars, and 'groomed' researchers away from 'dangerous' themes.

The course, then, was designed as an intervention and a necessary disruption of that tendency, and as a way to give students the chance to seriously engage with knowledges largely produced from the perspective of those traditionally constructed as 'the problem' or 'subjects of study' in dominant knowledge. This was and is in the interest of students and student-groups who, like their peers at other Danish universities, increasingly engage in endeavours to decolonise the curriculum and the university. The course was at the same time an act of love to the students who enter into academia searching for strong answers to strong questions (Santos 2014), only to find a horizon of possibility radically narrowed down by the purview and telos of dominant knowledge. Within this narrow horizon of possibility, fundamental issues pertaining to problems of racism, climate change, patriarchy, coloniality, etc. are set aside or addressed from a western knowledge framework. Finally, the course was also a way to bring together scholars at the university with a close relationship to the Global South, who find ourselves marginalised in different departments with little time or possibility to engage in the exchange of ideas and debate. In this sense, then, the book can be seen as having been born out of resistance to and struggle against the white, westernised university from the Global South—in Denmark. Indeed, while most contributors to this volume are 'Southern' by birth, ancestry, legacy, and socio-historical position, all engage thinking *with* the Global South. In a larger framework, the volume contributes to the current critical knowledge debates in Denmark and the world by engaging transmodernising, pluriversalising, decolonising, queering, and/or posthumanising thinking and practice.

Indisciplined transdisciplinarity

Addressing localised global problems spanning from informal transporta-tion in Mexico City (Finck Carrales) to water in the Himalayas (Poddar), this anthology reflects the pluriverse in its breaking out from the idea of a universal way of knowing and being in the world. Instead, it offers a multi-plicity of pathways of knowledges and practices that address current problems co-produced by the dominant Western colonial onto-epistemic outset. This means that even if transdisciplinary, the book must also be understood as 'indisciplined':

> To indiscipline means unleashing the boundaries of the social sciences that surround the production and distribution of knowledge, and the 'ontological regions' of the social, the political and the economic. It implies the recognition of other forms of knowledge, particularly the local knowledges produced from the colonial difference, and the dialogic crisscrossings and flows that can occur between them and the discipli-nary knowledges. It also proposes to break with the modernist trends of the social sciences that divide and alienate the subject and the object of knowledge, thus rethinking the dialogic relationship between subject and structure.
>
> (Castro-Gómez et al. 2002: 13–14, authors' translation)

Indisciplining entails questioning fundamental categories or notions like 'gender' and 'race' (Ghavami Kivi), 'individual' (Suárez-Krabbe), 'health' (Singla, Elbek and Fernando), and 'the border' (Mesbah and Asilgarajevs) without thereby implying that such categories are unreal. Instead, it implies questioning them, examining them in their complex historical embeddedness in a hard-lived global order reigned by a colonial structure which constantly shifts, or re-accommodates itself in order to reproduce itself. While such a global order is hard lived, the chapters of this book also point to ways in which something else is emerging or can emerge concomitantly with its decline. As such, the chapters reflect how the global order also shapes (trans)locally lived experiences of navigating a complex reality of both-ands, neither-nors, and not-yets. Such interstitial, transitional moments are addressed from different levels and perspectives. Carney and Timsina explore the different global and local complexities of Nepalese international students' lived experience in Denmark while Singla, Elbek and Fernando reflect upon concrete practices and methodologies to decolonise global health promotion. Finck Carrales applies a Dusselian transmodern philosophy of science to concrete settings of transport planning in Mexico City, while Poddar thinks with and along water scales in the Himalayas. Ghavami Kivi reclaims the decolonising power of the witch in a historically Western-dominated setting while Mesbah and Asilgarajevs address EU border-policies with Turkey as being simultaneously physical, geopolitical, and epistemic. Finally, Suárez-Krabbe engages in thinking through the existential dimensions of the decolonisation of the coloniser. All chapters engage such complexities in critical and purposeful ways towards change rooted in the diverse places, time-spaces, contexts, and beings they involve.

Additionally, the book's transdisciplinarity involves chapters that engage disciplines/perspectives such as posthumanism, feminism, decolonial perspectives, postcolonial studies, sociology, philosophy, etc. Each chapter also reflects different faces, so to speak, of global problems such as climate change (Poddar), the border regimes (Mesbah and Asilgarajevs), the structural-existential dimensions of decolonisation (Suárez-Krabbe), racism and patriarchy (Ghavami Kivi), globalisation vis-a-vis education (Carney and Timsina), global health (Singla, Elbek and Fernando), and urban transport services and local inequalities (Finck Carrales). Such problems are re-evaluated and re-studied from perspectives beyond the Western, dominant Eurocentric. By this, hegemonic dogmas are left aside or bracketed in order to give way to 'other' ecologies of knowledge (Santos 2014) and practice (see Fiddian-Qasmiyeh and Daley 2019). Such efforts involve the making constructive ways of relating to each other as mutually interdependent human and other than human beings (see Klein and Morreo 2019).

In methodological terms, the contributions compiled in this book found their own way through conceptions and methodologies according to their foci. This means that Western scientific approaches have undergone creolization (Gordon 2012), or are used as an argumentative tool instead of as a scientific instruction (see Dussel 2016). By referring to and reflecting on water, Poddar draws upon object-oriented ontology and speculative realism; Singla, Elbek and Fernando, and Finck Carrales address colonial mindsets (Ateljevic 2013) in global health and urban transportation planning, respectively; Mesbah and Asilgarajevs, and Carney and Timsina tackle and seek to push beyond abyssal thinking (Santos 2014); Ghavami Kivi and Finck Carrales utilise situated knowledges (see Haraway 1991); and Suárez-Krabbe employs relinking and existentialism (Gordon 1995; 1999) as a way to nurture the pluriverse. Even though the chapters draw upon and contribute to decolonial or postcolonial approaches, they also differ in terms of argumentative positions and propositions. All align with the idea of challenging the Western status quo, for instance, from a problem-oriented perspective (Singla, Elbek and Fernando), a critical feminist-queer(ing) perspective (Ghavami Kivi) and/or solutions-finding (Finck Carrales).

By providing transdisciplinary thinking and perspectives within the social sciences and humanities, the book additionally contributes to decolonising higher education. Other than looking at specific contexts with Southern 'lenses' as Fiddian-Qasmiyeh and Daley (2019) propose, each contribution engages ontologies, worldviews, practices, and epistemologies, which do not need any Western curator. This can involve a redefinition of questions, problems, and practices that account for the contextual epistemological 'needs' of the Global South as 'autonomous thought' (Escobar, 2017). In other words, we situate our positions when studying a phenomenon: speaking *with* our context, rather than *for* or in representation of our context or 'moving from a posture of "studying about" to "thinking with"' (Walsh 2018, 28). 'How to write, think, and act in ways that work to dismantle the structures of privilege and the modern/colonial matrices of power (of which privilege is part)' (Ibid, 21). Hence, for instance, theory as knowledge is context-situated and, at the same time, marked by modernity. Examples of these efforts can be seen in the political Constitutions of Bolivia (2009) and Ecuador (2008) (Latin-American Constitutionalism), which—albeit primarily only on paper—have incorporated and considered the Andean indigenous conceptions of *Pachamama* ('mother universe') and *sumak kawsay* and *suma qamaña* (living well) in order to provide nature with rights. In the case of Bolivia, nature is considered as a subject (Pinto Calaça et al. 2018). The incorporation of ancestral Global Southern ideas and worldviews into legal frameworks historically embedded in Western ontology (of the human-nature divide, for instance) opens the door to the incorporation of practices beyond such ontological frameworks, and have the potential to challenge capitalist and neoliberal logics. Additionally, these biocentric political practices aimed to respect the cycles of nature based on a non-developmentalist and non-exploitative rationality with nature (see Cornejo Puschner 2020) can potentially influence other Global South nations or even those of the Global North in order to alleviate the current global environmental crisis.

Global South in the North, Global North in the South

The conceptual separation between the Global South and the Global North can sometimes lead to the misunderstanding that the South is a homogeneous geographical unity, which is the counterpart of an equally homogeneous

Global North. This is a misconception. Countries usually attached to the notion of the Global South do in fact contain both the North, and many 'Souths'. The North and the South, as we use them here refer to different positions in the historically constituted unequal power relations between peoples. As such, the North is to be found in the Global South, notably among historical settler-colonial elites, and the South is to be found in the Global North, often but not exclusively embodied in the indigenous peoples, the (non-white) migrant, the refugee, and the colonial subject/citizen (Ndlovu-Gatsheni 2020). The Global South is also found in the knowledges and practices that were destroyed, subalternised and/or largely forgotten inside the Northern and Southern geographical territories (see Kudsk 2020, Ghavami Kivi this volume, Maxwell 2020, Suárez-Krabbe this volume, Federici 2004). This heterogeneity is indeed an important outset for this book's purpose.

However, we find it equally important to acknowledge that most peoples in the geographical south (together with some people in the north) share several *burdens* related to unequal power relations involving contemporary and past colonialisms and imperialisms (see Dussel 2012). Conversely, many people in the geographical north (together with a few elites in the south), have shared the *privileges* of such burdens. Such burdens include financial and economic terrorism (Valencia 2010), culture tropicalisation, appropriation and exotification (Lucia 2020), extractivism of knowledges and resources (Maxwell 2020; Tuck and Yang 2014), and science destruction and dismissal (Alvares and Faruqi 2012; Frizzo Bragato and Gordon 2017). In order to redress the losses of past and current colonial times, it is urgent to engage with and apply Global Southern worldviews (see Fiddian-Qasmiyeh and Daley 2019). The aim of such endeavours is to move towards power equality between the Global South and North. Such a utopic transition can be interpreted in many ways. One could be the pluriversality of Southern worldviews acknowledged by both the South and North until reaching epistemological equality (see Dussel, 2016; Escobar, 2017), or cognitive justice (Santos 2014).

Worldviews of many ancient and contemporary cultures are embedded in the Global South, which mirrors their historical contexts. This plurality of knowledges of the Global South can align/ally/associate with the Global North from an interdependent acknowledgment; unveiling and exposing the historical epistemic violence of modernity, as 'Western culture, with its obvious "Occidentalism," has positioned all other cultures as primitive, premodern, traditional, [...] underdeveloped[,] [...] unworthy, insignificant, unimportant, and useless' (Dussel 2012, 39–42). This entails searching and finding different connections and correlations between the Global South and Global North in order to disrupt universal ideas and achieve pluriversality. That is close to the South-South decolonial relationality/*vincularidad* that Mignolo and Walsh (2018, 1) pose although, in these terms, between the North and South '[...] in search of balance and harmony of life in the planet'.

Towards our transmodern pluriverse

The proposals that stem from this book are not 'alternatives' to Western ideas but rather these are contextual and independent without 'obeying' Western worldviews. That is, for example, different from the conception of postdevelopment in part as an 'alternative' to development for the Global South proposed in Klein and Morreo (2019). In the context of the westernised university, this critique relates to the idea that the university can be decolonised only by diversifying the academic staff, texts, and courses. While the incorporation and inclusion of knowledge and academic staff from the Global South in the Western university is apposite, such measures are steps in a much wider process, namely that of radically transforming the university to reflect, engage in, and nurture pluriversality. In other words, the idea is not to replace one knowledge with another—indeed, that would be to replicate the logics of the dominant system. Rather, pluriversality involves processes of co-creation, remembering, dreaming, walking, and doing together with the vast diversity of theories and perspectives that stem from and in the Global South. As an institution, the westernised university is part of the continuous reproduction of global (neo-)colonial power, but it also increasingly houses staff and students from the Global South. In this way, and as mentioned at the beginning of this introduction, the westernised university is undergoing a significant tectonic shake that is contributing to a de-universalisation that could be seen as a transition from uni- to pluriversity. Hence, Global South scientific approaches and perspectives should not be seen as alternatives to dominant Western Eurocentric knowledge, but as knowledges that nourish academic curriculums, fundamentally changing the university, which would then no longer revolve around a limited set of theories and perspectives based on the same onto-epistemic understandings of the world. Instead, it would engage, in theory and in practice, with and from the different places, time-spaces and ways of being and living that are actually found on Earth.

With regard to the epistemological aspect of such an idea, there is an increasing realisation that researchers from the Global South simultaneously critique or decolonise Western thinking while also applying decolonial and postcolonial approaches in the Global South to contextually address the problems we face and coming up with our own solutions. This is achieved by working in community and collaboratively designing, implementing and working with theories, methodologies and methods that stem from and are based on philosophies and practices in the Global South (see, for instance, Clair 2003; Kimmerer 2018; Leyva 2015; Suárez-Krabbe 2011; Tuhiwai Smith 1999; Paris and Winn 2014). Invoking Audre Lorde, Mignolo and Walsh (2018) formulate it as follows: 'If "another world is possible", it cannot be built with the conceptual tools inherited from the Renaissance and the Enlightenment. It cannot be built with the master's tools' (p. 7). In this context it is important to remember that knowledge and systematic thinking did not emerge with modernity, but has always been a human condition, and that people around the world did not stop thinking the moment they were colonised. Even though the conditions changed, people continued thinking about their problems, and the possible ways to move beyond them. When many Southern knowledges address the problem of colonialism, racism, imperialism, patriarchy, and capitalism it is not coincidental, but due to the fact that these became globalised with the European colonial endeavour since 1492.

To place Southern thinking in the past, or to expect it not to address these globalised problems as they localise in their specific settings, is to buy in to the inherited, dominant hierarchy of knowledge that gives Northern ideas and values monopoly over problem-framing and solutions-finding. Some higher education places, such as the Mexican *Universidad de la Tierra* and *Universidad Autónoma Chapingo* apply contextualised ancestral knowledge and are organised around principles of knowledge-sharing. Several examples exist of indigenous peoples or nations having elementary schooling grounded on the land, and including ancestral knowledges and teaching methods. Among these are some Mexican elementary schools of the Yucatán Peninsula that teach Mayan mathematical methods, and some elementary schools in Sierra Nevada de Santa Marta in Colombia, where a fundamental part of the curriculum has been developed on the basis of the teachings of Mother World. The discussions pertaining to the 'Africanisation' of knowledge in Africa (see, for instance, Cross and Ndofirepi 2017a; 2017b), as well as those contained within the vast plethora of ideas within Africana thought (see, for instance, Bogues 2003; Henry 2000), Islamic thought (see, for instance, Saeed 2017) and other Southern perspectives (Alvares and Faruqui 2012; Donald 2009; Graham 1999; Seth 2011), also attest to the fact that knowledge, practice and change go together, and often reflect the problems that peoples, and other beings around the world face. To say that they are all saying the same is to display either an intellectual arrogance or an intellectual laziness to actually engage these traditions of thought and learn about and from their heterogeneity. It is considering this vastness of the knowledges about our worlds that Dussel highlights the importance of a non-hierarchical acknowledgement and valuation of knowledges towards a transmodern culture 'that assumes the positive moments of Modernity', entailing an intercultural dialogue that 'needs to set out from a place-other than a mere dialogue between the learned experts of the academic or institutionally-dominant worlds' (Dussel 2012, 43).

Moreover, based on the comprehension of decolonial 'thinking-doing, and doing-thinking' (Mignolo and Walsh 2018, 9) as something that 'delinks, that undoes the unified—and universalising—centrality of the West as the world and that begins to push other questions, other reflections, other considerations, and other understandings' (Walsh 2018, 19), Southern thinkers often purposefully engage in breaking with the dominant Western understanding that knowledge construction, and even intelligence, is an individual matter which takes place and flourishes behind a desk. Instead, they work from the acknowledgment that knowledge construction is always a social activity that has practical, tangible effects—also in the future. The focal point for such science-making moves away from the false ideas of 'objectivity' and 'neutrality' towards considerations about what kind of world(s) we actually contribute to making with our research. The contributions contained in this book make inroads towards this end.

Under this prism, this book advocates for what Jaeger (2018) explains about the so-called 'ontological turn':

> Rather than leading to a mere plurality (and relativism) of worldviews, the ontological turn asserts that practical (and simultaneously conceptual) connections between cultural representations (including human and non-human thought and sensation) and nature (or materiality in general) entail an actual, that is, ontological multiplication of worlds.
>
> (Jaeger 2018, 228)

That multiplication of worlds involves a mutual understanding between the worldviews of the Global South and Global North. In academic practice, that can be achieved via consequent and open philosophical dialogues with the purpose of opening the scientific canon in order to multiply possibilities of conceptualising or diagnosing problems and finding solutions. Mark Jackson supports this idea by outlining 'good bridges' because 'building dialogue is about recognising and committing to difference [...] to appreciate different intellectual positions for their strengths and weaknesses, while also endeavouring new possibilities' (Jackson 2018, 5). In this way, the global transmodern future is possible as the understandings and representations of reality and its problematisation multiply. While the structural problems of racism, capitalism, patriarchy, colonialism, capitalism, and the depredation of nature are shared, they localise differently, and can be addressed, resisted, and changed in many different ways, while still being relevant to one another across these differences. As previously suggested with the case of the Bolivian and Ecuadorian constitutions, one important transmodern outcome of these is arguably the mutual nourishing that sharing understandings of how to overcome contextual social and environmental issues can produce. The solutions to these problems are particular but not exempted from being important contributions to think about watersheds in other contexts, perhaps especially in Western ones.

While the above-mentioned attacks on theories and perspectives such as those represented in this book are serious and powerful attempts at silencing the thinking practices towards another world, they must also be understood contextually, that is, as part of struggles to defend the current global system that kills and destroys. Concretely, they are struggles that defend the modern-colonial death project (Organizaciones Indígenas de Colombia 2004; Suárez-Krabbe 2016). As the chapters in this volume attest, such defensive struggles are not only a matter of the extreme right. Rather, they are the very fruits of the dominant Western ways of thinking and being that can only take place at the expense of all Earth-beings that do not bow, or comply with such destructive practices. In this sense, in addressing what world(s) we create with our research, the texts in this book while representing a heterogeneity of ideas and realities, all partake in offensive struggles against the death project, and move towards transmodern-pluriversal ways of being by contributing to opening up the horizon of possibility and the purview upon ourselves, each other, and the problems we face.

Notes

1. Authors' translation: *[...] even entire peoples can be corrupted, as when the population of the Empire is silent, looks the other way, before the immolation of innocent peoples [...]*
2. Authors' translation: *We do not believe in one unified direction for another world. But we believe that there is a strength in insisting that our various struggles can only be won if we meet each other and work together. For the power that threatens our existence is inextricably linked. We can count on each other.*

References

Alvares, Claude and Saad Saleem Faruqi. 2012. *Decolonizing the University. The Emerging Quest for non-Eurocentric Paradigms*. Pulau Pinang: Penerbit Universiti Malaysia and Citizens International.

Arce, José and Julia Suárez-Krabbe. 2018. "Racism, Global Apartheid and Disobedient Mobilities: The Politics of Detention and Deportation in Europe and Denmark." *KULT: Racism in Denmark* 15, 107–127. http://postkolonial.dk/wp-content/uploads/2017/09/11_Julia-og-Jose_We-are-here-because-you-were-there_final.pdf.

Ateljevic, Irena. 2013. "Transmodernity: Integrating Perspectives on Societal Evolution." *Futures* 47, 38–48.

Barreto, José-Manuel (ed.). 2013. *Human Rights from a Third World Perspective. Critique, History and International Law*. Newcastle: Cambridge Scholars Publishing.

Bhambra, Gurminder, Dalia Gebrial and Kerem Nişancıoğlu, (eds.). 2018. *Decolonizing the University*. London: Pluto Press.

Bogues, Anthony. 2003. *Black Heretics, Black Prophets: Radical Political Intellectuals*. New York and London: Routledge Series on Africana Thought.

Boidin, Capucine, James Cohen and Ramon Grosfoguel. 2012. "Decolonizing the University, Practicing Pluriversity. Special issue, Human Architecture." *Journal of the Sociology of Self-Knowledge* 10, no. 1. https://www.okcir.com/product/decolonizing-the-university-practicing-pluriversity/.

Castro-Gómez, Santiago, Freya Schiwy and Catherine Walsh. 2002. "Introducción." In *Indisciplinar las ciencias sociales. Geopolíticas del conocimiento y colonialidad del poder. Perspectivas desde lo Andino*, edited by Catherine Walsh, Freya Schiwy and Santiago Castro-Gómez. Quito: UASB/Abya Yala.

Clair, Robin. P. (ed.). 2003. Expressions of Ethnography: Novel Approaches to Qualitative Methods.

Cornejo Puschner, S. 2020. "Descolonizar la naturaleza: preguntas, tensiones, contradicciones y utopías." *Revista De La Academia* 30, 149–174.

Cross, Michael and Amasa Ndofirepi (eds.). 2017a. Knowledge and Change in African Universities: Volume 1–Current Debates. doi: 10.1007/978-94-6300-842-6.

Cross, Michael and Amasa Ndofirepi. 2017b. Knowledge and Change in African Universities: Volume 2 – Re-Imagining the Terrain. doi: 10.1007/978-94-6300-845-7.

Cupples, Julie and Ramón Grosfoguel (eds.). 2019. *Unsettling Eurocentrism in the Westernized University*. London and New York: Routledge.

Dhawan, Nikita (ed.). 2014. *Decolonizing the Enlightenment. Transnational Justice, Human Rights and Democracy in a Postcolonial World*. Berlin: Barbara Budrich Editors.

Donald, Dwayne. T. 2009. "Forts, Curriculum, and Indigenous Métissage: Imagining Decolonization of Aboriginal-Canadian Relations in Educational Contexts." *First Nations Perspectives* 2, no. 1, 1–24.

Dussel, Enrique. 2006. 20 tesis de política, México, Siglo XXI/Centro de Cooperación Regional para la Educación de Adultos en América Latina y el Caribe.

Dussel, Enrique. 2012. "Transmodernity and Interculturality: An Interpretation from the Perspective of Philosophy of Liberation." *Transmodernity: Journal of Peripheral Cultural Production of the Luso-Hispanic World*. 1, no. 3: 28–59.

Dussel, Enrique. 2016. *Filosofías del sur. Descolonización y transmodernidad*. México: Akal/Inter Pares.

Escobar, Arturo. 2017. *Designs for the Pluriverse. Radical Interdependence, Autonomy, and the Making of Worlds*. Durham and London: Duke University Press.

Federici, Silvia. 2004. *Caliban and the Witch*. New York: Autonomedia.

Feminist Autonomous Centre for Research. 2021. *Open letter against the targeting of our colleagues*. Athens, March 14, 2021. https://docs.google.com/forms/d/e/1FAIpQLSc1ch6XCPRqFc5lF1OjoxIPmgFT50c_70oORYrBRrxgrdaHww/viewform?fbclid=IwAR3Zi0uRqs8JDEuqCJoD10Ua89kTljL1arpRTVRk06H1SVqoG-qKU8f5rfI&gxids=7628.

Fiddian-Qasmiyeh, Elena and Patricia Daley (eds.). 2019. *Routledge Handbook of South-South Relations*. London: Routledge.

Freedom of Movements Research Collective. 2018. *'Stop Killing Us Slowly'. A Research Report on the Motivation Enhancement Measures and the Criminalization of Rejected Asylum Seekers in Denmark*. Freedom of Movements Research Collective, Copenhagen. http://refugees.dk/media/1757/stop-killing-us_uk.pdf.

Frizzo Bragato, Fernanda and Lewis R. Gordon (eds.). 2017. *Geopolitics and Decolonization: Perspectives from the Global South*. London and New York: Rowman & Littlefield International.

Gordon, Jane Anna. 2012. Creolizing as the Transdisciplinary Alternative to Intellectual Legitimacy on the Model of the "Normal Scientific" Community, *Quaderna*, permanent URL: http://quaderna.org/?p=415.

Gordon, Lewis. 1995. *Fanon and the Crisis of European Man. An Essay on Philosophy and the Human Sciences*. New York and London: Routledge.

Gordon, Lewis. 1999. *Bad Faith and Antiblack Racism*. New York: Humanity Books.

Gordon, Lewis. 2011. "Shifting the Geography of Reason in an Age of Disciplinary Decadence." *Transmodernity: Journal of Peripheral Cultural Production of the Luso-Hispanic World* 1, no. 2. https://escholarship.org/uc/item/218618vj.

Gordon, Lewis. 2020. "What to Do in Our Struggle to Breathe: Fanon's Relevance in Our Time of Multiple Pandemics." Online Lecture Delivered at the Caribbean Philosophical Association's Fanon at 95 event, July 1–20, 2020. July 13, 2020. Last accessed December 10, 2020. https://www.facebook.com/CaribPhil/videos/fanon-at-95-july-13th-celebration/593647091339732/.

Graham, Mary. 1999. "Some Thoughts about the Philosophical Underpinnings of Aboriginal Worldviews." *Worldviews: Environment, Culture, Religion* 3, 105–118.

Grosfoguel, Ramón, Roberto Hernández and Ernesto Rosén Velázquez (eds.). 2016. *Decolonizing the Westernized University: Interventions in Philosophy of Education from Within and Without*. Lanham, Boulder, New York, London: Lexington Books.

Haraway, Donna J. 1991. *Simians, Cyborgs, and Women. The Reinveintion of Nature*. London: Free Association Books.

Hassani, Amani. 2019. "Islamophobia in Denmark: National Report 2019." In *European Islamophobia Report 2019*, edited by Enes Bayraklı and Farid Hafez. Istanbul: SETA, 2020.

Henry, Paget. 2000. *Caliban's Reason: Introducing Afro-Caribbean Philosophy*. Routledge series on Africana Thought.

Jackson, Mark (ed.). 2018. *Coloniality, Ontology, and the Question of the Posthuman*. London and New York: Routledge.

Jaeger, Hans-Martin. 2018. "Political Ontology and International Relations: Politics, Self-Estrangement, and Void Universalism in a Pluriverse." In *Coloniality, Ontology, and the Question of the Posthuman*, edited by Mark Jackson. London and New York: Routledge.

Kimmerer, Robin Wall. 2018. *Braiding Sweetgrass. Indigenous Wisdom, Scientific Knowledge and the Teaching of Plants*. Minneapolis: Milkweed Editions.

Klein, Elise, and Carlos Eduardo Morreo. 2019. *Postdevelopment in Practice: Alternatives, Economies, Ontologies*. London: Routledge.

Kothari, Ashish, Ariel Salleh, Arturo Escobar, Federico Demaria and Alberto Acosta (eds.). 2019. *Pluriverse: A Post-Development Dictionary*. New Delhi: Tulika Books.

Kudsk, Dea. 2020. "At genordne tænkningen. En studie i forbindelser mellem undertrykkelse af viden og destruktion af verden." Masters thesis, Cultural Encounters. Department of Communication and Arts, Roskilde University.

Leyva, Xochitl (ed.). 2015. *Prácticas otras de conocimiento(s). Entre crisis, entre guerras. I, II, II*. Cooperativa Editorial Retos. San Cristóbal de Las Casas, México.

Lucia, Amanda J. 2020. *White Utopias. The Religious Exoticism of Transformational Festivals*. Oakland: University of California Press.

Marronage. 2017. Marronage er kollektiv modstand. *Marronage, 2*. Forlaget Nemo.

Maxwell, Abby. 2020. "On Witches, Shrooms, and Sourdough: A Critical Reimagining of the White Settler Relationship to Land." *Journal of International Women's Studies* 21, no. 7: 8–22. https://vc.bridgew.edu/jiws/vol21/iss7/2.

Mignolo, Walter. 2007. "Delinking. The Rhetoric of Modernity, the Logic of Coloniality and the Grammar of De-coloniality." *Cultural Studies* 21, no. 2: 449–514.

Mignolo, Walter and Alvina Hoffmann. 2017. Interview – Walter Mignolo/Part 2: Key Concepts. *E-International Relations*. January 21, 2017. https://www.e-ir.info/2017/01/21/interview-walter-mignolopart-2-key-concepts/

Morton, Jeffrey. S. (2000). *The International Law Commission of the United Nations*. Columbia: University of South Carolina Press.

Nanni, Lucion and Enzo Traverso 2021. "Islamophobia, 'Islamo-leftism', (post)fascism." *Interview*, Versonbook blogs. https://www.versobooks.com/blogs/5018-islamophobia-islamo-leftism-post-fascism.

Ndlovu-Gatsheni, Sabelo. 2020. *Decolonization, Development and Knowledge in Africa. Turning over a New Leaf.* New York and London: Routledge.

Organizaciones Indígenas de Colombia. 2004. *Propuesta política y de acción de los pueblos indígenas. Minga por la vida, la justicia, la alegría, la autonomía y la libertad y movilización contra el proyecto de muerte y por un plan de vida de los pueblos*. Last accessed December 10, 2020. http://anterior.nasaacin.org/index.php/2010/06/04/propuesta-politica-de-los-pueblos/.

Paris, Django and Maisha T. Winn (eds.). 2014. *Humanizing Research. Decolonizing Qualitative Inquiry With Youth and Communities*. Los Angeles, London, New Delhi, Singapore, Washington DC: Sage.

Pinto Calaça, Irene Zasimowicz; Carneiro de Freitas, Patrícia Jorge; da Silva, Sérgio Augusto; Maluf, Fabiano. 2018. La naturaleza como sujeto de derechos: análisis bioético de las Constituciones de Ecuador y Bolivia. *Revista Latinoamericana de Bioética*, 18, no. 1: 155–172.

Rutazibwa, Olivia U. and Robbie Shilliam (eds.). 2018. *Routledge Handbook of Postcolonial Politics*. London: Routledge.

Saeed, Abdullah. 2017. *Islamic Thought. An Introduction*. London and New York: Routledge.

Santos, Boaventura de Sousa. 2014. *Epistemologies of the South. Justice against Epistemicide*. London and New York: Routledge.

Seth, Sanjay. 2011. "Travelling Theory: Western knowledge and its Indian object." *International Studies in Sociology of Education* 21, no. 49: 263–282.

Özcan, Sibel, and Zeynep Bangert. 2019. "Islamophobia in Denmark: National Report 2018." In *European Islamophobia Report 2018*, edited by Enes Bayraklı and Farid Hafez, 251–282. Istanbul: SETA. http://www.islamophobiaeurope.com/wp-content/uploads/2019/09/DENMARK.pdf.

Siddiqui, Usaid. 2021. "Muslim Austrian Academic Shares Tale of Gunpoint Raid." *Al Jazeera*, March 4, 2021. https://www.aljazeera.com/news/2021/3/4/muslim-professor-reveals-raid-in-austria.

Speri, Alice. 2021. A Billionaire-funded Website with Ties to the Far Right is Trying to "Cancel" University Professors. *The Intercept*, April 10, 2021. https://theintercept.com/2021/04/10/campus-reform-koch-young-americans-for-freedom-leadership-institute/.

Suárez-Krabbe, Julia. 2011. En la realidad. Hacia metodologías de investigación descoloniales. *Tabula Rasa*, 14, 183–204 Universidad Colegio Mayor de Cundinamarca Bogotá, Colombia. https://www.redalyc.org/pdf/396/39622094008.pdf.

Suárez-Krabbe, Julia. 2016. *Race, Rights and Rebels: Alternatives to Human Rights and Development from the Global South*. London: Rowman & Littlefield International.

Suárez-Krabbe, Julia and A. Lindberg. 2019. "Enforcing Apartheid? The Politics of "Intolerability" in the Danish Migration and Integration Regimes." *Migration and Society. Advances in Research* 2, no. 1, 90–97.

Suárez-Krabbe, Julia. Forthcoming. Relinking as healing. On crisis, the collapse of whiteness and the possibilities of decolonization. *Globalizations*, Special Issue: "Our bodies breathe resistance: COVID19 Stories from/in the Margins".

Tuck, Eve and K. Wayne Yang. 2014. "R-Words: Refusing Research." In *Humanizing Research. Decolonizing Qualitative Inquiry with Youth and Communities*, edited by Paris, Django and Maisha T. Winn. Los Angeles, London, New Delhi, Singapore, Washington DC: Sage.

Tuhiwai Smith, Linda. 1999. *Decolonizing Methodologies: Research and Indigenous Peoples*. London: New York Zed Books.

Valencia, Sayak. 2010. *Capitalismo Gore*. Spain: Melusina.

Mignolo, Walter and Catherine Walsh. 2018. "Introduction." In *On Decoloniality Concepts, Analytics, Praxis*, edited by C. Walsh and W. Mignolo. Duke University Press.

Walsh, Catherine. 2018. "The Decolonial For Resurgences, Shifts, and Movements." In *On Decoloniality Concepts, Analytics, Praxis*, edited by C. Walsh and W. Mignolo. New York: Duke University Press.

Woons, Marc and Sebastian Weier (eds.). 2017. "Border Thinking and the Experiential Epistemologies of International Relations." (OA here: https://www.e-ir.info/2017/06/02/border-thinking-and-the-experiential-epistemologies-of-international-relations/)

1 Globalisation in theory and practice

Negotiating belonging in Danish higher education

Stephen Carney and Nitya Nanda Timsina

Introduction

Dilu's dream

One day, I shall leave Denmark and return to my home of Pokhara in the hills of Nepal. I'll build a hotel for the many tourists. In the morning, I'll sit on the roof-top, amongst my fruit trees, soaking up the early sun. I'll enjoy my oranges and look out across the lake as the clouds drift by. I'll be an entrepreneur with many employees. I can direct them by mobile phone. When she is grown, my daughter can remain here in Copenhagen to live the life I was denied. The dream of education brought me here but it has become a hard reality. The only cure is to go home where I belong.

In search of the west

Furba and Prem, students at a local business college in Copenhagen, decided to go to the beach. It was the hottest day of the year, but it was not the cool water that enticed them. Both were studying in large classes comprised completely of Nepali students like themselves. They saw Danes on the trains and city streets but knew few. The college is full of young men and women who have sacrificed all to journey across the world in search for something more. Today, Furba and Prem wanted to see Danish bodies, hopefully uncovered. Pure and white, white and pure, just as they were on the TV screen back home, glowing with the future. They came upon a scantily clad young woman who started a discussion. Why had they come to the beach in jeans and woollen sweaters? Why were Nepalese men so short? Being so poor, it must be strange to come to the beach and relax, right? She seemed friendly but also too curious and too close to uncovering the truth behind appearances. In any event, it turned out that she was a tourist from the Netherlands. The two men moved on, now more self-conscious of their complete incapacity to belong.

★★★

DOI: 10.4324/9781003172413-2

This chapter examines the theory framing of globalisation debates and explores how these framings shape life choices and experiences, especially amongst international students from the so-called Global South. It argues that concepts shape realities and that these have consequences, especially in contexts that have been historically ordered and marginalised by dominant ways of seeing. Globalisation literature deals with processes, ideologies, and a multitude of practices with complex forms and effects. Some of this, especially early writing, is unashamedly positive (Beck 1992; Friedman 2005). Later writing appears more measured, occasionally sceptical (Axford 2013). Importantly, the theorising of globalisation—especially within the Western academy—has practical and personal effects that shape lives and the possibilities for achieving individual fulfilment and societal recognition. It this sense, we can view theory as *creating* reality. Many international students find themselves caught between competing worldviews, with profound consequences.

Theoretically, we consider the phenomenon of globalisation in terms of unequal power relations that are manifest in contemporary flows of information, technology, people, and ideas but, also, in terms of its relationship to the legitimation and spread of a certain understanding of Western modernity that is nevertheless received differently and reworked in relation to other world views. Globalisation is thus many things but, from this perspective, we might best understand it as a particular type of 'abyssal thinking' that makes certain values and life forms visible while obscuring, distorting, or eliminating others. To illustrate the power of theory to shape the imagination and human pathways, we draw on the life experiences of Nepali higher education students in Denmark, most of them are in search of something that remains painfully elusive. 'Northern' thinking plays its part in ordering social relations and must be approached critically if we are to imagine and fashion alternative realities that do justice to the richness of global experience and the mutual respect that, it is claimed by policy makers at least, must be at the heart of cross-border collaboration.

We begin by outlining some ways in which globalisation has been presented in the Western academic literature before examining some of its pernicious consequences. Our approach to the study of global phenomena is to view the world as interconnected but where flows of values, culture, symbols, commitments, and understandings are uneven. International students enter such global flows with much at stake. Some national groups move to countries and programmes that connect to deep histories of interaction, for example, earlier colonial pathways where language and cultural expectations create a certain space for belonging and identification. In the Nepali case, one could think of flows of peoples and practices to the United Kingdom. Others find their way to new locales such as Australia, which competes for fee-paying students and requires low-wage labour for its services sector. Other international students must pioneer untested routes into countries and contexts that have, at best, ambiguous or ambivalent attitudes and support structures for embracing immigrants (including international students) from different, especially minority, backgrounds.

Studying mobility is by no means easy. Concepts such as nation, state, and culture are destabilised by global flows, meaning that one must attend to multiple forces that influence the production of subjectivity. Methodological nationalism (in particular, the attempt to view phenomena through the singular lens of the nation state and its cultural narratives) has its limits. Mobile young people reflect local *and* cosmopolitan beliefs and values, as well as desires to belong to particular communities, as well as to resist incorporation into the identity projects of others. Finding one's way home—wherever that might be—is thus a project of struggle and discovery not easily reduced to the dichotomies insisted upon by political groups eager to split, divide, and re-establish old divisions. In our time, the international student may have multiple belongings and identity projects, feeling at home everywhere and nowhere simultaneously. As they meet inclusive and restrictive forces, identities take form as reflections on belonging and exile.

We focus on Nepali students in Denmark who reflect a very particular segment of the international student community in this country. As a relatively new group of entrants to Denmark, they are both welcomed by educational providers keen to marketise their institutions but met with various degrees of paternalism by a general public that identifies Nepal with the international development discourse of poverty and the romance of unspoiled natural/primitive beauty. Most recently, an increasingly populist right-wing nationalism links international students from the Global South to historically devalued groups from the so-called Middle East, who are defined primarily by race and religion. Nepalis in Denmark thus find themselves in a complex field marked by processes of inclusion *and* marginalisation. However, these processes are not only the product of some base Danish racism but, rather, central to the deepening of a particular form of global capitalism that builds upon and extends historic processes of exploitation, repackaging them as opportunities for social mobility through investments in human capital and personal growth from integration into the cosmopolitan world view.

In methodological terms, we mirror the fragmented nature of global connections with an empirical strategy that has prioritised the collection of loosely organised stories and the display of those 'productive entanglements' (Thrift 2008, p. viii) that complicate the life of the exile. Many of these are heart-rending accounts of struggle and disenchantment and our aim has been to honour those with as little re-working as possible. Many Nepalis speak of their predicament by drawing on episodes from home, from the educational field, from relationships, and in terms of their fears and desires. They are not necessarily coherent if one takes the perspective of rational Western science with its demands or order, clarity, and academic purpose. If only life was as simple as science! Instead, we present our 'data' in partial glimpses that attempt to do some justice to the complex webs of associations raised as international students try to find existential shelter. In so doing, we hope to manifest as method something of the tumultuous task of finding a space in a familiar/unfamiliar place and how globalising education projects are implicated in those.

Exploring globalisation research

The trope of globalisation has been worked continuously since a surge of interest in the wake of the collapse of the Berlin Wall and transformation of the Soviet Empire. Notwithstanding this profound moment in history, the globalisation of ideas, peoples, and things is not new. One way to nuance contemporary discussions is to consider the distinctions between processes of *globalisation*, ideologies of *globalism* and *globality* as experience (Axford 2013).

As process, globalisation research explores structural change and new dynamics between nations, institutions, and peoples. Much early writing in this genre was openly celebratory, with the promise of a 'new kind of economy, a new kind of global order, a new kind of society, and a new kind of personal life' (Beck 1992, p. 2). How one responds to this vision depends very much on one's notion of globalism. One approach to globalism, strongly liberal capitalist, envisages a 'flat earth' of movement and integration, and thus opportunity and wealth (Friedman 2005). Others take this optimism into the political sphere to suggest a new era of justice where the coming 'Empire' emerges from the 'twilight of modern sovereignty'. Unlike our earlier imperialist era, this new world 'establishes no territorial centre of power and does not rely on fixed boundaries or barriers. It is a decentred and deterritorialising apparatus of rule that progressively incorporates the entire global realm within its open, expanding frontiers' (Hardt and Negri 2000, p. xiii). Even though this vision looks towards a new democratic relation between peoples, others have viewed it as naive. For example, '... what is generally called globalization is a "vast social field" in which hegemonic or dominant social groups, states, interests, and ideologies collide with counter-hegemonic or subordinate social groups, states, interests, and ideologies on a world scale' (Santos 2006, p. 393). This perspective suggests a brutal process of legitimising inequality and domination through the deception of language and the complicity of the global elite. Globalisation is not only a process that brings the world together but also one that creates and normalises differences and inequalities. It connects places but also bypasses them. From this perspective, the subject is unavoidably caught in global circuits of capitalism that seek out those in most need and insert them into complex systems of production, consumption, and exploitation (Thrift 2008). Globalisation thus includes moments of exclusion and abjection (Ferguson 1999) but also forms of active participation that may well result in limited possibilities and new miseries.

Perhaps there is no pre-conceived possibility here and that we should instead focus on how the world is experienced. Globality, then, is an orientation that is cautious of 'false universals', 'unreflexive aggregation' and the prioritising of economic modes of thought. It is also aware of an ever-present Euro-centrism in thought (Nienass 2013, p. 534). Globality requires a respect for affects and consequences, as well as an awareness that place is no longer a static container of history and culture but an ongoing event where new prac-tices are made (Massey 2004).

What makes our current era so challenging to analyse is the new role of the imagination in framing how we experience the world. A spatial approach to the social sciences requires that we view phenomena in motion, respecting their historicity but remaining attuned to the de-territorialised dimensions of life. Arjun Appadurai (1996) theorised the global economy in *cultural* terms, where global dynamics have a 'complex', 'overlapping', and 'disjunctive' character. His notion of 'scapes' (ethno, media, techno, finance, and ideo) suggests an end to simpler centre/periphery distinctions and, instead, a new era of interconnectivity but, importantly, one based on the visual, the emo-tive, and the ideal. One can certainly imagine international students within Appadurai's conceptual model of a World joined by the problematic com-bination of abstract ideas and overheated desires. Here, the global subject becomes a product of a utopian language of possibility as well as a deeper psychological fear of exclusion and abandonment.

Much of the imagery of the global comes from the very concepts used to describe it and these are far from neutral. One analysis asks us to consider globalisation theory as a form of 'Northern' knowledge (Lizardo and Strand 2009), where the domination of classical concepts from the European history of ideas limits worldviews to those that are familiar to the refined Western reader. As a consequence, it is not surprising that much of the literature con-cerned with the 'global condition' emanating from Northern academic centres is defined—even if only implicitly—by Anglo-American concerns related to issues such as class (Marx), state (Weber), and social solidarity (Durkheim). Rather than address the Global South on its own terms, we find instead the 'North' colonising the 'South' and the 'South' itself largely appropriated as a research object and site for amelioration. Such writing produces 'reality' in part by producing silences. Even when academics from the North attempt to understand and describe new global dynamics from the perspective of the South, they do so via recourse to Northern thinking. When, for example, Hardt and Negri (2000) outline the contours of a coming new era, they do so via the social thought of Lukács, Benjamin, Adorno, Wittgenstein, Foucault, and Deleuze. Strangely, there is no place for Gandhi or Fanon and indeed, no voice from beyond the dominant European order. Connell (2007, p. 379) notes that:

> Neither Bauman nor Beck, nor Robinson nor Kellner nor Sassen, refers to nonmetropolitan social thought when presenting theories of globalization. Nor does Robertson, despite his career in development studies …. At the end of *Runaway World* Giddens helps the reader with an

annotated reading. All 51 books mentioned are published in the metropole, and only one of them centrally concerns a nonmetropolitan point of view. Giddens's account simply does not address nonmetropolitan thought about globalization. It is a striking fact that this body of writing, while insisting on the global scope of social processes and the irreversible interplay of cultures, *almost never* cites nonmetropolitan thinkers and *almost never* builds on social theory formulated outside the metropole.

Such writing is dangerous because it obscures the pervasive ways in which discourse frames what can be thought and written and assumes that we can operate from the position that the world is framed by a singular global *episteme* (Connell 2018). Indigenous traditions of thought and, even experience, are erased and, by virtue of the rules of good science, treated as provincial, ill-disciplined, beyond generalisation, and thus of limited value in the important debates of the North. However, the experience of the world that is whitewashed by such writing does not vanish from the world. The Global South is more than an 'object' of globalisation theory, or 'data mine for sociology'. Rather, the people of these worlds are 'producers' of globalisation theory (Connell 2007, p. 381).

This line of thinking is well known by scholars who have identified an 'abyssal' line between legitimate (scientific) thought and it is invisible, disruptive other (Santos 2016, p. 118). Here, the denial of 'nonmetropolitan experience' becomes essential to what *can* be thought and said:

> Modern knowledge and modern law represent the most accomplished manifestations of abyssal thinking. They account for the two major global lines of modern times, which, though being different and operating differently, are mutually interdependent. Each creates a subsystem of visible and invisible distinctions in such a way that the invisible ones become the foundation of the visible ones. In the field of knowledge, abyssal thinking consists in granting to modern science the monopoly of the universal distinction between true and false, to the detriment of two alternative bodies of knowledge: philosophy and theology…that cannot be fitted into any of these ways of knowing. On the other side of the line, there is no real knowledge; there are beliefs, opinions, intuitions, and subjective understandings, which, at the most, may become objects or raw materials for scientific inquiry.
>
> (Santos 2016, p. 119)

A 'southern knowledge' alternative to such genocidal thought starts from the acceptance that 'the understanding of the world by far exceeds the West's understanding of the world' (Santos 2016, p. 164). A Southern approach respects ecologies of knowledge, where scientific and analytical approaches are treated as perspectives with no greater claim to universality than the multitude of 'performative, literary, magical, and non-human forms of understanding' that also comprise the human experience (Carney and Madsen 2021, forthcoming). Nonmetropolitan social thought can thus come *from* the South but also be an awareness of such plurality *within* Northern scholarship. It requires new languages, conceptual repertoires, and perspectives, as well as a diligence to remain alert to the creeping tendency in science to draw us back above the line, where only realist, critical, universal, binary, and reductionist thought is viewed with ease. Such tendencies are not only embodied with the language and practices of Western science. They occur through publishing patterns and priorities, resource allocation and forms of self-governance, where southern thinkers (and in our case, many international students) are tamed by systems of patronage, training, reward, and early socialisation to reject their own experience, histories, and knowledges in order to find acceptance into the pathways offered by global flows of knowledge and power.

These are not abstract tendencies but social practices that remake minds and bodies. Here, the argument is that when young, mobile, students wish to continue their studies; they do so from a diet of reading and thought that demonizes and dismisses the history, achievements and potentials of their own worlds while praising as exemplary and necessary the worldview of others. One manifestation of this is the very desire to study in the Northern metropole itself, with the certainty that one's own intellectual ballast is, by definition, inadequate. Nepali students are most certainly familiar with this phenomenon, having received a constant stream of messages through formal schooling, media representations, and political discourse, of the inadequacy of the village (read Nepal) in relation to the city (read the cosmopolitan centre) (Pigg 1992; Shrestha 1995). Once engaged in academic work in the Metropole, many see no connection between their scholarly identity and ethnicity (Kim 2021). It is hardly surprising, then, that a huge disconnect occurs when students from the Global South meet systems of inclusion/exclusion in European universities and society. It must be an aim of a decolonising sociology to understand these processes from the perspectives of participants and to do so with a 'mosaic' approach to epistemology (Connell 2018, p. 404) that gets beyond one point of view or dominant framing of the problem. One starting point for an alternative approach to exploring the burgeoning field of international student mobility is to work with the desire, ambition, pain, and regret of participants who find themselves in a 'discontinuous state of being' (Said 2000, p. 141). For us, that means working with 'untamed stories' (Alinje 2019) that defy simple categorisation and ordering by the themes established in the existing literature. One can *draw* upon the literature of the metropole without being beholden to it.

Glimpsing globalisation as/in practice

International student mobility has become a major factor in the development of higher education philosophies and systems worldwide. The establishment of the European Higher Education Area (EHEA) frames the Danish context, where almost 50 nation states collaborate in ways that make possible deep levels of cross-national and cross-system mobility. This includes the establishment of a common degree framework within the EHEA, an extensive system of support for student and staff exchange, as well as a focus on European-wide themes within research funding bodies. As Erfurth (2020) notes, mobility has played a major role in the development of not only of synergies across educational programmes and their content, but also in shaping new European identities. The EHEA links to a European policy ambition to promote the region as a leading knowledge society with international student mobility viewed as an essential part of this vision. Nepali students understand this vision as an invitation to join the emerging European space, itself one manifestation of early, optimistic iterations of the globalisation heuristic. They comprise a major segment of incoming, non-EU students to Denmark, with many undertaking two-year vocationally oriented courses before attempting to settle in the country or move into the university sector. The costs of full-fee education in Denmark are considerable, as are the challenges of obtaining additional visas that enable permanent (or at least extended) residence and the possibility of settling families and career paths. The Danish immigrant service acknowledges that foreign students may have a need to supplement their finances with paid employment alongside their studies. International student engagement in Denmark is thus conceived as taking place alongside active participation in the labour market. One could argue that the Danish economy actually *depends upon* a large number of low-salary, non-permanent, and semi-skilled workers. Here, one can speak of integration, but only of a certain type. Danish universities have also indicated a desire to enhance their involvement with international students, which can be seen by the prioritising of improved institutional structures to support their lives and learning. Finally, one can also surmise that after 30 years of intense 'development' assistance by the Danish Government to multiple social sectors in Nepal (framed by the language of 'partnership'), there exists a genuine commitment to engaging with the country by including Nepalis in Denmark itself.

These potentially positive factors must be balanced against a number of other imperatives that have worked against the broader acceptance of many student groups from the Global South. The Danish labour market remains highly regulated, reducing employer incentives to take on unknown workers at the same rates of pay as Danish counterparts. The national economy—like society—is heavily shaped by the Danish language, which is notoriously difficult to master, certainly within the timeframe of most student visas. While universities wish to recruit new student groups, the funding model in place within the university sector is not dependent on the recruitment of fee-paying students. Indeed, few Danish universities actively seek to recruit students from the Global South. If one examines Danish university recruiting materials, as far as these exist, it would seem that the ideal international student is the cosmopolitan/Northern, student of science or technology who, after graduation, would seamlessly enter such fields within Danish industry, where issues of language and culture may be less determining factors than they are for those from the humanities and field such as management, teaching, and social work. Finally, but of growing significance, Denmark has been marked by an extremely negative period of policy development related to immigration, where debates about refugee numbers, structural under-employment within communities of migrants and their descendants, the perception of growing EU regulation and, even, international terrorism, are brought together in ongoing processes of othering and belonging (e.g. Warren 2019). The notion of 'Danishness', while unhelpful to such debates, and which in any event continues to be notoriously difficult to define, is commonly used in discussions about the Nation's future cultural direction, with most of the major political parties now accepting a right-wing immigration agenda that, 20 years ago, was viewed as extreme. Indeed, the Danish political agenda in this regard has become less noticeable in a post-Brexit Europe framed by resurgence nationalisms. International students from the Global South find themselves caught in the cross hairs of such debates.

The struggle for success

Furba came to Denmark from an extended period as a 'contractor' to a US security firm in Kabul, Afghanistan, a country (or 'situation') that had provided respite for the many young Nepalis mired in the unemployment and abjection of home. At the end of his contract, he made a brief trip home and then flew to Denmark with nothing but a few hundred dollars, considerable trepidation and a small serving of hope. He had worked as a cook for the Americans but now wanted more. In the entrepreneurial hothouse of Kathmandu, he devised a plan that had served many others well. The Danish government offered work visas to the spouses of international students. It was not hard to find a Nepali woman who had been granted a study place in the nascent Danish market for foreign students and it was surprisingly easy to get the agreement of her family for a marriage of ultimate convenience. The arrangement would be that he could live and work in Denmark as her spouse but that they would not live together or have any meaningful contact. For his part, he was obliged to finance her tuition fees by working. 'Problems always have many solutions', he said with a wisdom that betrayed sorrow and regret.

★★★

Furba and his close friend Prem work in a South Asian restaurant in Aarhus. The owner had another four restaurants in the city and elsewhere, employing scores of young immigrants trying to cover their living expenses. Many are 'international' students, meaning those from beyond the European Union and Common Economic Area. Contrary to a political discourse in Denmark which paints them as a drain on the Country's resources, it is they who provide an economic life-line to the many Danish colleges fighting for fee-paying students and the legitimacy and recognition that comes from 'feeding' economic value into the Danish university system. This is an enormous market in which Danish institutions compete for students both at home and abroad.

For the owner—known as 'the Boss'—money is also the barometer of a good life. He lives in confidence, deeply attached to his possessions, which includes cars, apartments, and a constantly changing background of Danish girlfriends. Furba adds to his earnings as a cleaner at the Boss's luxury apartment located in the most expensive part of the town. The Boss had used some small part of his recent profits to buy this dazzlingly address and fill it with antique furniture, carpets, enticing sofas, and beds. 'If only he would give me the kennel', cries Furba, 'I would sleep there more comfortably than I do in my dirty basement'. He prays every morning, 'O God, if you give me a better room, I promise I will build a temple for you!' Furba says that the Boss had moved away from Eastern European girlfriends to Danish women. He was sure that it was all part of a strategy that combined social positioning and revenge on his adopted home and its hard identity politics directed against Muslims and immigrants from his homeland. Furba's situation could not be more different. He had no contact with his 'paper wife' beyond the regular transfer of income to cover her fees. When the longing became too great, he would spend his few savings with the Eastern Europeans down near the bar district. They were always polite, never judged him and only took what seemed reasonable.

One afternoon, the Boss approached Prem and Furba and asked one of them to take on a full-time role in the business. Furba was definitely uninterested and indicated that Prem might want to take the job. 'But I have to go to university', Prem replied. The Boss bellowed: 'What's the use of your international degree my friend? I came to Denmark to study medicine. I would have become a doctor if I went to university. But look what I have become now?' He paused as Prem lowered his head in submission. Prem's part-time work consisted of opening and closing the downtown restaurant. By day, he went to university. The Boss asked him to consider the offer seriously and went off to solve other pressing challenges. Prem remained in a swamp of misery. For all his bluster, the Boss was right. His education would eventually be complete and even now, he knew it to be useless. He would never get the white-collar job in this country that his friends and family were expecting. The Danes hardly saw him at all and their labour markets were only interested in his muscle and submission. 'The reward for my degree in Denmark will be a monthly salary about the same as what the Boss pays the cook at the downtown restaurant. It's nothing. Doesn't this look like failure to you? Where will it end?'

Dilu finances her husband's education in Denmark

Dilu was married on her 19th birthday so that she could join her husband in Denmark who had been accepted for a graduate study programme. She now works full-time in a department store in Copenhagen, paying her husband's tuition fees and covering their monthly expenses. Finding it impossible to get a job through the Danish employment services, she turned to her own network of women and eventually took the job at the store. She is known for her dedication and hard work and is a key member of the support staff. Dilu now works with her Pakistani friends and a team of Indians, Kurds, Turks, Bengalis, Arabs, Macedonians, Bosnians, Serbians, and Africans. This shadow world is fundamental to the robust and competitive Danish national economy. 'Who would have imagined' says Dilu, 'we are the invisible caste that carries an ignorant nation! The world calls Denmark the happiest country and the reason can be found here. Here, we have the whole UN working for minimum wage so that Danes can have the cheap products they need to find the contentment they deserve'.

Dilu is now learning Urdu. 'Danish would be a waste of time as I don't interact with the local people. At home, I used to hear Urdu only on the television. Now, this is like my own tongue'. Urdu was not a choice. Her employer, a Danish citizen with a strong Pakistani identity, demanded that she learn a more useful language. He is a strong character and not especially enamoured with his employees or, for that matter, his customers. The diversity of nations in the store leads to constant dispute about politics, identity, and the eternal struggle of belonging. One Indian is particularly difficult. Dilu says that he barks out orders at her constantly, yelling at her to run faster, to find better produce, and to offer a lower price. He is disbelieving that she does not speak Hindi and complains to the manger that she will only speak Danish to him. After their last exchange, the Indian customer demanded that Dilu be fired for insubordinate. This gave the store manager much enjoyment: the chance to antagonise the Indian was delicious: 'But she is our best employee sir', he offered in a conciliatory tone. For Dilu, placed in a foreign land with the sole purpose of providing security to her new husband, Denmark is a world away. 'I try to remain a good Hindu and to learn the ways of the Danish society but I must also be loyal to Pakistan, hostile to India and friendly to the young Danish customers who are rude and aggressive. It's like being at the frontline of a conflict I don't understand and can't escape'. Education was to be a way not only to master such complexity but to transcend it: the entry point for a different outlook on the world. 'I thought that when my husband completed his studies, he would find a good professional job and we could begin a different life. Now, he is a cook in a restaurant in Copenhagen and our hopes are dashed'.

Coming home

While many of international students arrive in Denmark with their living and financial arrangements in place, others, especially those from South Asia, struggle desperately just to survive and do so with little security or personal well-being. Sharmila came to Denmark as a 'paper wife'. She saw no future at home and only the bright lights of an inclusive cosmopolitan centre in

Denmark. She met a young man in Nepal through mutual friends and they agreed to a secret contract: they would marry officially and travel together as husband and wife. In a reverse of the arrangement made by Furba, it was now the husband who had the student visa and Sharmila who would get the spousal visa and chance to live the good life. The price was that she would earn enough money in Denmark to pay 50% of her husband's university fees. She travelled to Europe alone and arrived in Copenhagen with no place to stay and no friends or family to support her during the early weeks. Her husband was sharing a small room with four other student-friends and there was no space. Her first night was a blur of sitting and standing in the crowded space, afraid to wake the many inhabitants as they slept. Eventually, she made contact with a Nepali family who took her in for a month and after some weeks, she found a cleaning job in a large hotel, thanks to their Nepali network. In her first week, she was threatened with dismissal for not being able to clean each of her 15 rooms within 15 minutes. At that point, she phoned her father in Nepal to say that she would be coming home as soon as possible. Europe was so difficult, cold, and foreign and she could not see a future there. 'Europe' was not what they had talked about with such clarity and confidence back in Nepal.

During these early days, Sharmila drew upon her upbringing in a large family to negotiate around the crowed house where she slept on the floor next to her married hosts, careful not to disturb their sleep by her movements. Her daily routine continued in Nepali with the highlight being the evening meal that provided a tangible connection to her old life back home. Once her job was secure and income could be assured, her 'paper husband' exercised his claim for payment. She had been offered easier money by a hotel guest who, sensing her precarious status, offered to provide her a room in his luxurious apartment elsewhere in the Country in turn for sexual favours. Here, she knew her limits: 'If I allow my body to be used in such ways it hardly matters to stay alive. It is within my body that my soul resides and it is this I must protect and take home to my country and family'. Until she could afford the return journey, her focus was on maintaining employment in order to qualify for an extension to her visa. This would become urgent when her husband's studies (and thus period of residency) ceased. 'It is strange, my life is structured by his ability to complete his program. It is like, I too, am an international student'.

Between belonging and abjection

International education, like other fields of human endeavour, is structured by intersectional dynamics that are far from fair or just. In the phantasmagoria of global flows, study abroad can be a pathway to something desired and realised but, also, its opposite. It can embody what Laurent Berlant (2011) calls, *cruel optimism*. Through its cluster of promises, international education both obscures the forces that make it possible while reinforcing their most pernicious effects. In our study here, we most certainly find students who had realised the many possibilities attached to international education but also those who were seemingly deluded into believing that they were entering a level playing field of opportunity and merit. Students from South Asia seemed particularly trapped in the realm of limited possibility.

Prior to coming to Denmark, Ganga was viewed by his village community as an extraordinary talent, one who would go far academically. His parents thought he would become a great engineer. That vision led to studies in Nepal and Denmark and, ultimately, long-term unemployment and disillusionment. 'I feel as though the hatred towards foreigners that is championed by right-wing nationalists began to show its ugly head at about the same time as South Asians began to show interest in a life in Scandinavia. Ganga thinks that he is a victim of a 'nasty racial profiling' that is becoming widespread across Europe. A mathematics and science teacher in an international school in Kathmandu before coming to Denmark, he completed a master's degree from the Copenhagen Business School before settling in the City as a permanent resident. Unable to find work appropriate to his degree, he now drives a public bus. It is not for want of trying: 'I can't remember how many applications I wrote but nobody called me back. What does it mean?' One answer may lie in a recent study that suggests that even Danish politicians are more likely to ignore written requests from their constituents if their names are recognizably connected to ethnic minority groups (Berlingske 2021). Still, Ganga aims to remain in Denmark for the benefit of his young daughter who receives high quality public schooling and free healthcare. He says that she often remarks that he teaches mathematics much better than her 'real' teachers.

★★★

Shyam completed a master's degree in social entrepreneurship and management and mentions this only to appease the shame of the long-term unemployment that followed. After years of struggle, he has taken a part-time job as a cook in one of the many downtown restaurants in Copenhagen. He keeps in touch with his classmates by distance, usually via their Facebook posts, which seem to present nothing but opportunity and success. His teenage sons ask when he will find such 'prestige' jobs. 'They must be out there, waiting to be grabbed', says one. However, his preoccupation is to remain in the world of appearances, exploring ways to present a successful veneer to family and friends back home. 'I don't know how long this can continue and how I can move on. I am trapped here in a world of mirrors. Going home is not an option'.

★★★

While many international students from Nepal despair in their inability to develop meaningful relationships with 'the white Danes', others found a measure of self-understanding and growth in this forced estrangement from both homes. For Sanjay, a sense of belonging came not from the maintenance of relations with family and friends back in Nepal but, rather, through his attachments to the Nepali cultural network in Copenhagen. This is a source of inspiration to remember home but also to understand it in relation to the Western modernity that was always held up as its superior form. He explains:

> It's true that our village life is poor in money terms and I myself viewed home and my world through the lens of poverty. But that's a limited view. There is richness in the seasons, understanding the rhythm of the daily life of a cow, of how the bananas ripen just when they are needed and how, at home, we live with an optimism that helps cast dark thoughts from the mind and look to the future.

Life in Copenhagen had helped him realise that there are many measures of progress. The village seemed to get its 'light' from the glow of the city and that was a promise as well as a trap. 'Yes, we came to forget the joy of the village and home. Now, I have nostalgia for home only because I am here and so far away. The village comes to life again with all its colours, smells and sounds'. Perhaps it requires life in this grey busy, self-important city for home to show itself?

Conclusion

Late in the afternoon, a group of Nepalis climbed to a little mound next to Husum Torv with the *pāncē bājā* (a collection of traditional musical instruments) and thought about their villages back home. Kiran played the *naraśin'gā* (horn), Yam played the *tyāmkō* (drum), Hari played the *chyālī* (cymbals), and Suresh played the *sanai* (pipe). Their music captured as ecstatic longing for another world; one often missed but rarely thought of as a feasible destination after journeys of such pain and dislocation. On the mound—a mere flick of dirt in comparison to their Himalayan homeland—Nepal could be conjured into existence, if only in tones and tunes that dispersed quickly into the cold wind and grey sky above. 'Maybe my family can hear this tune as it travels, maybe not', says Suresh.

★★★

When Sanjay initially considered undertaking a period of study abroad, he imagined not only a life in Europe, but the associated trappings of Western modernity: iPhones and tablets, Apple watches, high-speed internet, and streamed music. Now, he laments that with many of these desires satisfied, the longing for home grows stronger. 'Some days, I get stuck with images from my old life: the pumpkin flowers and creepers from my family home surrounded by the wild bees in the mustard fields, the evening chorus of frogs and the early morning songs of the sparrows in the guava trees'. Amongst the crowds of commuters and shoppers in Copenhagen, Sanjay feels alone, set apart and vulnerable. 'This is strange: At home we had to worry about the porcupines and wild boars in the cornfield. Here, we must be weary of the nationalists and fascists who look at us with such contempt'. Is this the idea of 'progress' and 'development' that we were so curious about back in the village? Another of his friends, now scratching out a living in the United States, wrote to him, explaining the gap between a cosmopolitan ideal and the longing for a way back home:

> Hi *da* [bro], I'm fine and hope all is well there. Yes, those dreams were lovely. I would say those dreams worked as a fuel to keep us moving in some direction, right? NYC is a great place to live, a true melting pot of the world. However, honestly *bhannu parda* [speaking], me and my wife now talk and dream about someday returning to Nepal. We talk to each other about those beautiful images of green hills and tea gardens! Wow! da I am just hoping this dream will again work as a fuel to drive us back!

Haunted by a past that continues to return and humiliated by a visible and tangible future that remains forever out of reach in Copenhagen, he finds ways to carry on. 'Maybe we'll go back, but to what? I can't just leave all that is good here like schools and health care and our freedom from all the obligations of home and the violence of politics. What is Nepal anyway? A dream? Like Denmark?' Perhaps cultural diversity and cosmopolitan attachment requires that such immigrants join its 'society of strangers' (Hall 2013, p. 6) and accept membership of the many 'black holes' of social exclusion that characterise global modernity (Morrow and Torres 2000, p. 49). The growing 'fear of small numbers' (Appadurai 2006a) redefines the experience and meaning of international student mobility for many on the borders of the new world order:

> The continuing reality is that globalisation as a new phase or new moment in the history and mobility of capital continues to produce all sorts of irregularities in the tectonics of political and social life. In one sense nothing has changed after September 11, 2001: the nation state is certainly only one among many other players, though it is evidently neither dying nor dead. There are many kinds of sovereignty, popular and transnational, surrounding the claims of the nation state. Some are progressive, others are criminal, some are separatist, others universalist. Also, citizenship is, now more than ever, a complex field and most national polities are occupied in part by people who are not full citizens, but are partial or marginal citizens. This is of course the political story in Europe today.
>
> (Appadurai 2006b, p. 170)

The Nepalis present in these pages navigate the unsettled borderland between the global matrix of majority vs. minority politics, establishing a form of belonging in Denmark that will do for now. Rather than settling the unsteady ground on which they stand, international education may well add to that instability, creating new thought, *and* 'ideocide' (Appadurai 2006a, p. xi), hope *and* fear, as well as home *and* exile.

References

Alinje, Rahul. 2019. "Examining Indian NCF 2009 Policy as an Assemblage: Subjectivity & Silent Spaces." Unpublished thesis of Doctor of Philosophy. Roskilde University.

Appadurai, Arjun. 1996. *Modernity at Large: Cultural Dimensions of Globalization.* Minneapolis and London: University of Minnesota Press.

Appadurai, Arjun. 2006a. *Fear of Small Numbers.* Durham: Duke University Press.

Appadurai, Arjun. 2006b. "The Right to Research." *Globalisation, Societies and Education* 4 no. 2: 167–177.

Axford, Barry. 2013. *Theories of Globalization.* Cambridge: Polity Press.

Beck, Ulrich. 1992. *Risk Society: Towards a New Modernity.* London: Sage.

Berlant, Lauren. 2011. *Cruel Optimism.* Durham, London: Duke University Press.

Berlingske, Tidende. 2021. Politikere vil hellere svare på e-mail fra 'Anne' end 'Fatima' (Politicians would rather reply to an e-mail from 'Anne' than 'Fatima'). Monday 1 February 20021. https://www.berlingske.dk/danmark/kommunale-politikere-vil-hellere-svare-paa-e-mail-fra-anne-end-fatima (accessed 8.3.21).

Carney, Stephen and Ulla Ambrosius Madsen. 2021. *Education in Radical Uncertainty: Transgressions in Theory and Method.* London: Bloomsbury.

Connell, Raewyn. 2007. "Northern Theory of Globalization." *Sociological Theory* 25 no. 4: 368–385.

Connell, Raewyn. 2018. "Decolonizing Sociology." *Contemporary Sociology* 47 no. 4: 399–407.

Erfurth, Marvin. 2020. "International Education Hubs in the Global Education Industry. Changing Policy and Governance in Higher Education." Unpublished thesis of Doctor of Philosophy. University of Münster.

Ferguson, James. 1999. *Expectations of Modernity: Myths and Meanings of Urban Life on the Zambian Copperbelt.* Berkeley, CA: University of California.

Friedman, Thomas. 2005. The World Is Flat: A Brief History of the Twenty-First Century, Farrar, Straus and Giroux. http://capitolreader.com/bonus/The%20World%20Is%20Flat.pdf (accessed 6.11.18).

Hall, Susanne M. 2013. "The Politics of Belonging." *Identities: Global Studies in Culture and Power* 20 no. 1: 46–53.

Hardt, Michael and Antonio Negri. 2000. *Empire*, 47–69. Cambridge, MA: Harvard University Press.

Kim, Terri. 2021. "The Positional Identities of East Asian Mobile Academics in UK Higher Education: A Comparative Analysis of Internationalisation and Equality and Diversity." In *Identities and Education: Comparative Perspectives in Times of Crisis*, edited by E. Klerides and S. Carney, London: Bloomsbury.

Lizardo, Omar and Strand Michael. 2009. "Postmodernism and Globalization." *Protosociology* 26: 38–72.

Massey, Diane. 2004. *Space, Place and Gender.* Cambridge: Polity.

Morrow, Raymond. A. and Carlos A Torres, 2000. "The State, Globalization and Educational Policy." In *Globalization and Education, Critical Perspective*, edited by Nicholas C. Burbules and Carlos Alberto Torres, 27–56. New York: Routledge.

Nienass, Banjamin. 2013. "Performing the Global." *Globalizations* 10 no. 49: 533–538.

Pigg, Stacy. 1992. "Inventing Social Categories through Place: Social Representations and Development in Nepal." *Comparative Studies in Society and History* 34: 491–513.

Said, E. W. 2000. *Reflections on Exile and Other Essays*, 137–149. Cambridge: Harvard University Press.

Santos, Boaventura de Sousa. 2006. "Globalizations." *Theory, Culture & Society* 23 no. 2–3: 393–399.

Santos, Boaventura de Sousa. 2016. "Beyond Abyssal Thinking." Ed. Santos, Boaventura de Sousa. In *Epistemologies of the South: Justice against Epistemicide*, 118–135. London: Routledge.

Shrestha, Nanda. 1995. "Becoming a Development Category". In *The Power of Development*, edited by Jonathon Crush, 266–277. London: Routledge.

Thrift, Nigel. 2008. *Non-Representational Theory: Space, Politics, Affect*. London: Routledge.

Warren, S. 2019. "ReThinking the 'Problem' in Inquiry-based Pedagogies through Exemplarity and Worldriented." *Education Sciences* 9: 295.

2 Transmodern philosophy of science in the case of informal transportation in Mexico City

Local ontology and epistemology for transport planning

Juan Carlos Finck Carrales

Introduction

Countries that historically have suffered colonial settlements and/or 'developing' countries that are not located in the 'developed' Western Europe, Asia, and Anglo–America form the Global South. Today, colonization is still ongoing in many ways through supremacism between countries. Science production, acknowledgement, and validation from the Global North to the Global South are part of that global paradigm.

Based on that paradigm and after three years of investigating transport planning in Mexico City, in 2016, I planned a PhD research project. Its case study was an informal motorcycle cab service (MCS) in one peripheral neighbourhood of Tláhuac municipality in Mexico City. The service has provided local mobility to thousands of inhabitants for the last 20 years. This phenomenon also involves thousands of employment opportunities across the city. However, the MCSs have also caused an increase in air pollution, an informal–unequal economy, and even criminal activities.

The service is currently undergoing a formalisation process towards cancelling such negative externalities. However, historical governmental lack of attention and few scientific Western positivist studies on the phenomenon have not given satisfactory results in that process (see e.g. Farrés Delgado and Matarán Ruiz 2014). According to O'Donnell, Kramar, and Dyball (2013, 11), the 'positivist approach focuses on logical inferences from theoretical concepts (i.e. deduction) or a generalization from evidence (i.e. induction) [...]'. Researchers guide their studies from previous propositions that generalise/universalise reality and, consequently, research outcomes.

When I started this research, I had in mind to not reproduce those positivist scientific practices even though I wanted to make use of Western-manufactured theory. This might sound contradictory. Nevertheless, my aim was not to dismiss, discredit, or/and overlook Western knowledge but to break the traditions of the Western scientific validations and science guidelines application commonly practiced in Mexico. Mexican governments embrace such guidelines via Public Servants and their consultants with 'Westernised

DOI: 10.4324/9781003172413-3

mindsets' ('Westernised' stakeholders) (see Ateljevic 2013)—which is linked to what Quijano has termed 'the coloniality of power' (Farrés Delgado and Matarán Ruiz 2014; Quijano 1991). This elucidates and depicts a current Western ontological territorial occupation in the Global South (Escobar 2016). What I conceive as a 'Westernised mindset' aligns with what Madina Tlostanova (2017) describes as the coloniality of (ontological) design: '[...] a control and disciplining of our perception and interpretation of the world, of other human and nonhuman beings and things according to certain legitimized principles. It is a set of specific ontological, epistemic and axiological notions imposed forcefully onto the whole world [...]' (Tlostanova 2017, 3).

I wanted to approach the MCS in order to explore possibilities of formalisation and improvement beyond the current westernised mindsets in Mexico. The study provided transport planning, fine-tuned methodological, and conceptual outcomes, which the authorities of the city have considered for the MCS formalisation. These relate to a contextual and interpretative application and enriching of the social position concept (Bourdieu 1996; Larsen and Morrow 2009) and structural stories as a method of analysis (Freudendal-Pedersen 2009; 2018) under the transmodern philosophy of science. The aim was to acknowledge and, possibly, balance power relations between different social groups in the study zone. Furthermore, the recommendations for transport planning in Mexico City, which stemmed from these findings, rise above positivist Western elements. I summarise these as following: (1) I argue that governmental representatives should go to the field and take part in meetings with local actors to reshape their worldviews about a phenomenon via an embodied experience. (2) This involves being aware of the power balance between stakeholders, which challenges, and puts into perspective the 'Westernised mindsets' and world views of governments when addressing Global South contexts. (3) I recommend giving more responsibility to the municipalities when it comes to transport planning in order to balance governmental power and favour a contextual approach to planning. (4) Likewise, the Mobility Law of the city should provide autonomy to local transport cooperatives. (5) Finally, create punishments to patronising and intimidating activities, which affect those groups.

This chapter aims to highlight the outcomes of my MCS study in which I applied the *transmodernity* philosophy of science (Dussel 2012; 2016). My contribution with this research rests on an application of the transmodern approach beyond merely an abstract treatment in research. In doing so, I ground theory until reaching empirically based findings. Overall, the transmodern approach aims to create a South–South dialogue between philosophers until reaching a South–North multicultural dialogue as a 'pluriversal' utopia (a horizontal/non-hierarchical understanding and acknowledgement of many worldviews and realities in the world besides the Western one). Here, Global South researchers have the possibility of using re-defined/interpreted European knowledge as 'basis tools' and background that 'dialogue' with their Global South case studies, favouring situated knowledge. This involves providing an interpretive approach in order to glimpse different angles of the phenomenon beyond deterministic logics, universal propositions, theory testing/adaptation, rational-choice, time-space efficiency, etc.

What I mean with 'dialogue' refers to the acknowledgement and incorporation of 'peripheral knowledge' to my analysis coming from local stakeholders in the study zone: European culture (in the form of Western theory and interviews with 'Westernised' Mexican stakeholders) in 'dialogue' with Mexican culture (in the form of data as interviews with Mexican locals) (see Grosfoguel 2011). This is possible because as Ateljevic (2013) mentions by paraphrasing Dussel, '[...] (subconscious) Eurocentrism currently pervades all cultural arenas, European and non-European, which makes genuine multiculturalism and dialogue [...]' (Ateljevic 2013, 41) and creates a sort of relation(ality) between Western and Eastern ontologies (which I used to achieve my research aim) (see Escobar 2016).

In my research case, the importance of contextual and interpretative approaches rests on the aim of providing social sciences with the idea of producing transport planning framed under local ontologies and epistemologies. Ramon Grosfoguel (2011) argues that 'Subaltern epistemic perspectives are knowledge coming from below that produces a critical perspective of hegemonic knowledge in the power relations involved' (Grosfoguel 2011, 6). Differently, my findings do not work as alternatives to Western thinking nor merely involve a critique to hegemonic knowledge based on Western–Southern power relations (in my case study, Westernised actors–Local actors power relations). My findings rather entail a contextual and local epistemology and ontology. Ontology in the sense of the specific meanings, characteristics, and attributions that individuals give to a certain phenomenon (a specific world-view); whereas epistemology is perceived as the validation and recognition of knowledge and its process of production, which emerge from an ontology.

Therefore, in order to warn decolonial scholars, I would like to clarify that my research did not aim to decolonise per se transport planning in Mexico City but rather to create contextual knowledge for planning by interpreting/ re-defining and interweaving Western theory to data and a philosophy of science of the Global South. There are two reasons why I had to use Western theory in my research, which partly led me towards using the transmodern approach: (1) I do not have access to local and contemporary theory, which is historically based on my study zone worldview and hermeneutics (native myths shaping different facets of reality through language expressed in texts [see Alvesson and Sköldberg 2000; Dussel 2016]) in conceptions akin to transport planning and social position (Western concepts), as local indigenous knowledge has been lost and/or actively destroyed since colonial times. Furthermore, I do not speak the native languages. (2) In addition, Western theory in those fields is the only one to which I have access. This includes theory manufactured in the Global South, since it is, too, historically based on Western hermeneutics. Thereby, in my research, I did not use Western theory as a Western guideline that defines notions of local data (positivist approach) but rather as a Western background notion to be re-defined through a local data interpretation (transmodern approach) (see Grosfoguel 2011).

That is precisely the essence of my theory-data 'dialogue' stemming from my interpretation of the transmodern approach. Therefore, due to the lack of use of local theory that probably does not exist, my research did not reach a 'purified' decolonial outcome, since I do not offer a 'fundamental cognitive shift' in the field based on 'Indigenous epistemes' (Tlostanova, Thapar-Björkert, and Knobblock 2019, 293). In connection to that, by contradicting the transmodern approach with regard to the possibility of making use of 'positive'/'benevolent' elements of modernity in Global South research, Mignolo and Walsh (2018) argue that to reach decoloniality, we should '[...] delink from the theoretical tenets and conceptual instruments of Western thought'. (Mignolo and Walsh 2018, 7). Regardless of those dogmas within the decolonial field, I argue that my research outcomes demonstrate a local episte-mology and ontology of my case study phenomenon coming from addressing; acknowledging; and putting into perspective power relations in horizontal ways and to 'dialogue' power relations and knowledge between Western(ised) and non-Western(ised) social groups. This local ontology and epistemology shaped my previously mentioned transport planning proposals for the MCS.

In what follows, I first present the transmodern approach for Latin-American science. This is based on ideas beyond modernity and post-modernity for avoiding universal statements and for creating other epistemologies besides the Western. In relation to planning studies, I address the South-Eastern turn in planning as this criticises Western science monopolisation. I argue, how-ever, that it is necessary to do research that aims to go beyond acknowledging non-Western cultures. Therefore, it is also necessary to address ontologi-cal and epistemological aspects regarding Global South cases when doing research. Furthermore, I consider the possibilities of reaching transmoder-nity within transport studies. I provide examples of Global North research that uses interpretative approaches and is unclosed to new methodology and method creation and application, opening the doors for further transmo-dernity approaches within that field. Finally, based on interviews I carried out with key MCS actors in Mexico City (from which I interpreted struc-tural stories), I map worldviews, power relations and social positions between social groups, and I contextualise governmental planning and action related to the MCS phenomenon.

Transmodern approach in science for Latin-American studies

Going on a different path from the South-Eastern turn in planning

The last decade or so, has seen the start of a wave called the 'South-East turn in planning' created by mostly Western scientists. This approach is directed mainly at planning studies and criticises strongly the monopolisation of science by Western culture. Those studies propose a cultural and contextual turn for planning, avoiding universality (Watson 2016), acknowledging indigenous people's perspectives in Western countries (Porter 2006), producing concepts in place (Roy 2009a; 2009b), working with cultural identity to democratise communities (Yiftachel 2016; 2006), etc.

The above mentioned research has criticised modern universal ideas for knowledge production and have offered new perspectives outside Western guidelines. The authors stress planning as part of the main problems of colonial thinking after Europe and Anglo-America imposed modernity thinking globally. Hegemonic governments (United States of America and Europe) and international financial organisations have historically imposed planning in the Global South under deterministic 'development' logics and goals, resulting in socioeconomic and spatial disparities in those territories (Galland and Elinbaum 2018).

In my MCS research, I aimed to improve transport planning in Mexico City to understand and let its culture create its own knowledge (based on interviews). I thereby did not aim to provide a research process that 'speaks for' the Global South 'natives' as the post-colonial approach suggests (see Tlostanova, Thapar-Björkert, and Knobblock 2019). Instead, I aimed to have a project that 'speaks with' the 'natives' and vice versa. That aspect goes beyond a critique of modernity and post-modernity that still looks at the Global South with Global North 'eyes'. Even by doing fieldwork in the Global South, it is not sufficient to wear the 'Global South glasses' at an ontological and, even less, epistemological level. As Tlostanova (2017) clarifies, we need to reject and stop teaching and preaching Western universalist claims to people of the Global South. Thus, the dialogue North–South starts from the local knowledge.

To this end, it is important that I analyse the philosophical approach that works with cultural hermeneutics, epistemology, and ontology by going beyond modernity and post-modernity through what has recently been manufactured in Latin-America. This approach is the transmodernity project.

Enrique Dussel's transmodernity: beyond post-modernity

Over the last 50 years or so in Latin-America, a wave of philosophy of science has been developing that aims to dismiss the imposed worldviews of Western science inherited since the colonial periods of the 16th century as the unique and truly universal ones. This approach can be used in research processes by acknowledging, self-valuing, and working with South-Eastern cultures in hermeneutical, epistemological, and ontological ways. This project is called transmodernity and has the main aim of creating a South–South dialogue between philosophers from countries that have undergone coloniality. Afterwards, it should be possible to reach a South–North critical and multicultural dialogue, in which all philosophers, including Western thinkers, will converge and share science in a 'pluriversal' utopia (Dussel 2016).

Dussel (2016; 2012) points out that it is important to acknowledge that all philosophies are based on cultural-historical myths of certain civilisations from a hermeneutical focus and that currently there is a dominant one: the European (extended into the Anglo-American). This started with the colonization of Asia, America, and Africa, where Western philosophies dismissed and looked down on all philosophies that were not European and in which non-Europeans were considered as inferior 'sub-humans'.

Under that understanding, Dussel (2016), together with other scholars, started the thinking wave of *philosophy of liberation* at the end of the 1970s, which consisted of creating a critical dialogue of culture in the sense of critical theory. This philosophy questioned the modernity inheritance of Western science that was taking place mainly in Latin-America. Afterwards, while Dussel continued his studies for decades, his project evolved to something that he called transmodernity.

Transmodernity is a world project that goes beyond European and Anglo-American modernity, meaning that it is post-Western by avoiding the European science superiority of the last five centuries and by strengthening the forgotten philosophies of the 'other world'. Transmodernity also goes beyond post-modernity, since, the latter is partly a partial critique to modernity made by Western thinkers. The main aim of transmodernity is that scientists and thinkers whose homelands were 'victims' of colonization articulate a self-reflexive discourse and acknowledge their own history.

Transmodernity stresses the 'periphery' of the World as different from the 'centre' presented as Europe and Anglo-America, which aims to generate consensus between colonization 'victims' through solidarity. The project starts by 'fighting' for the acknowledgment of the 'others' (the non-Europeans/ Anglo-Americans) and recovering creativity through self-valuation. Furthermore, this necessitates a South-South philosophical dialogue, namely between the different world scientific 'victims' of the Western domain.

Therefore, the transmodernity project is an anti-domination one of cultural liberation through an intercultural dialogue. Thereby, 'trans' in this sense means beyond and before modernity, since transmodernity does not 'touch' modernity due to it aims to avoid it completely because modernity is historically based only on capitalism, eurocentrism, and colonialism. However, the transmodernity project does not try to set aside Western thinking completely, but rather to get the important outcomes of it as selected tools that can entail benefits (mainly technological) in a final South-North open dialogue, which will be critical and, therefore, 'pluriversal' (see also Grosfoguel 2011). This paradigm will be possible only if both Southern and Northern philosophers are reflexive/open and acknowledge the value and meanings of the 'other' philosophies.

In relation to my research, the application of this approach involved an analytical discussion between Western theory and contextual data in the Global South. The approach also entails considering the different ontologies and worldviews of Westernised and non-Westernised people, which was carried out through interviews in Mexico City. These ontologies and world-views differentiate between stakeholders depending on their geographical positions, social classes, and even skin colours. Some are closer to the culture of the study zone and others are closer to European.

Moreover, the following quote from Grosfoguel's (2011) work helps to situate concretely my transmodern approach:

> Dussel's transmodernity would be equivalent to 'diversality as a universal project' which is a result of 'critical border thinking', 'critical diasporic thinking' or 'critical thinking from the margins' as an epistemic intervention from the diverse subalterns locations. Subaltern epistemologies could provide [...] a 'diversality' of responses to the problems of modernity leading to 'transmodernity'.
>
> (Grosfoguel 2011, 27)

Thereby, by applying the transmodern philosophy of science, the European theory I used in my research was merely a basis tool and background for re-defining, developing, and creating a scientific Northern-Southern theoretical 'dialogue', rather than mirroring or fitting it into my study context. This entailed 'another' knowledge directed and focused specifically on my study zone in order to attend to its issues and, in turn, interpret its 'reality'.

Contextual approach for research analysis and knowledge production

My conception of contextuality relates to my interpretation of what Donna Haraway (1991) calls *situated knowledges* from feminism (local knowledges instead of world systems/universality). Haraway uses situated knowledges to describe and criticise Western modern science domination over the rest of the world. She explains that the majority of science has been created under positivist closed epistemology guidelines overlooking 'non-capitalist-white-patriarchal' events and worldviews, even with their attempts of avoiding bias. Haraway therefore argues that modern science is 'rhetoric', since only 'powerful' people can reach the 'manufacturing' of knowledge in their own language.

Tlostanova (2017) explains that Haraway's situated knowledges concept differs from that of the pluriversal, as she believes that the former promotes a universal good for all instead of a constant communication of countless abstract options. I nevertheless argue that it is important to acknowledge and use contextuality to create pluriversal 'dialogues'. Under this line, the analytical innovation of transmodernity entails enriching theory in order to create new scientific knowledge (including theory manufactured by South-Eastern scholars under the same epistemological guidelines as European 'branch offices'). Haraway (1991, 187) supports this idea by calling to translate 'knowledges among very different—and power-differentiated—communities' around the world through 'solidarity in politics and shared conversations in epistemology' (*ibid.*, 191). In this sense, my research analysis consists of modelling reality through methods and, afterwards, interplaying their outcomes with theory.

In connection with this approach, current research within urban studies are partly based on making and searching for comparisons between other cities in the world in order to determine their 'development' (see Beaverstock, Smith, and Taylor 1999; Sassen 2005). For instance, for decades the government of Mexico City has spent vast amounts of money on receiving 'consultation' from different European and USA researchers and agencies with this purpose in mind. One example was Gehl Architects' (2009) consultation from 2007 to 2009 regarding the potential of a bike lane system in Mexico City, hoping that it was similar to the one in Copenhagen as an 'inspiration' to 'transfer' just because the latter is efficient.

That is an attempt of implementing Western urban planning hegemony in Latin-America, which, according to Farrés Delgado and Matarán Ruiz (2014), reflects and involves territorial coloniality of power ('Westernised' Public Servants), knowledge (positivist approach and Western theory) and being (urbanity over other types of living areas). As a response to that, Delgado and Ruiz suggest considering and using new urban epistemologies that reflect the identity and values of local indigenous peoples or citizens. I transfer such an understanding to the transport planning field.

Towards reaching transmodernity within transport planning studies

In order to reach transmodernity in science production, research needs to be open, reflexive, and, above all, critical of Western modernity and post-modernity thinking. Thereby, transmodernity can take place in different research disciplines, such as transport planning studies.

Currently, within academic research, I could not find an intersection between Dussel's transmodernity and transport planning studies. The theoretical approaches that I can relate to my case study research have been manufactured mainly in Europe. As explained in the Introduction, this situation is a result of not having access to my native lands indigenous languages from which I could create and start a hermeneutical approach of my study (see Solares 2009), towards an 'independent' theorisation of the phenomenon. This situation meant that I had to use theory based on European worldviews.

However, some Western thinkers had already indirectly opened the door to a possible transmodern approach regarding transport planning. Their interpretative approaches inspired me even though their works are not related to the informal transportation field. These researchers suggest the use of different contextual aspects, such as new vocabulary; people's culture; emerging topics; the creation of new questions, theories, and methodologies; interdisciplinary and integral focuses; among others (see Adey 2010; Büscher, Sheller, and Tyfield 2016; Sheller 2014). The authors' approaches acknowledge a possible epistemological change of a theoretical focus based on an interpretative methodology that goes beyond positivist Western normativity. Therefore, having a transmodern approach means creating open argumentations to fine-tuning theoretical conceptions, meaning the possibility of enriching Western concepts based on the analysis of my phenomenon.

Moreover, by using the interpretative and transmodern approaches together, I do not risk erasing the context, dismissing the colonial background (as 'Westernised mindsets'), which could favour overplaying Western theory over the contextual methodology and methods. In my case study, implementing interpretation without transmodernity therefore can still involve having Western theoretical propositions as unreflective adoption to some extent, which I want to avoid.

Social position and structural stories

The study of social positions and structural stories helped me with mapping the stakeholders' power relations and worldviews relative to the MCS phenomenon. Some of the MCS stakeholders of my study case belong to the neighbourhood and others to other areas of Mexico City. Bourdieu (1996) acknowledges that 'human beings' are social agents related to a social space and who occupy a place, so that focusing on their situations means focusing on their taken space. When those agents 'consume space', they are holding power from their occupied place, involving 'inherited' social hierarchies, and distances because their world positions depend on those of their ancestors passing from generation to generation (Larsen and Morrow 2009). Thereby, *social space* is an invisible set of relationships that form a definite distributional arrangement of agents and properties. The hierarchies and distances in the space make it symbolic, and its differences are visible in practices, possessed goods, and expressed opinions among agents. Each agent is in and speaks for his/her particular 'institution' as a representation of their occupied social space.

For example, Freudendal-Pedersen's (2018) analysis of city planning through storytelling within a reflexive methodology states that 'preconceptions about gender, place of living and education, blur the ability to hear the other stories that might be common despite different types and living conditions' (Freudendal-Pedersen 2018, 247). In relation to social class differentiations, in Mexico City, governments tend to belong to privileged social groups that keep a distance from excluded social groups. As a result, the government does not take proper care of issues related to underserved zones of the city. Thus, socially constructed characteristics of people can define the course of their ability to practice and conceive different levels and types of power. Freudendal-Pedersen thereby explains how an individualised understanding of autonomy is related to a specific society, namely, its time-space context in order to relate analysis of senses to people's storytelling for planning. Researchers can 'translate'/interpret people's storytelling as structural stories. According to Freudendal-Pedersen (2009), these are expressions of common stories within everyday life conversations. Structural stories can become 'common truths' for all members of a society as the substance of rationales that produce and reproduce actions.

By this means, stakeholders' social positions and structural stories helped me with seeing linkages between discourses and social process as differences in power and responsibilities. Finally, those 'levels of power' potentials need to be defined regarding the contextual and temporal situations of the stakeholders involved in a phenomenon.

Post-structural approach vs. transmodern approach

I would like to clarify that even though I used the social position concept (Bourdieu) and structural stories (Giddens) as a method in my analysis, this case study is not post-structural but rather transmodern. Ultimately, my research involves interpreting the theory (not testing it), as explained in the Introduction, considering the epistemic North-South power relations between social groups, to reach transmodern empirical findings for transport regulation and planning.

The post-structural approach would have not allowed reaching such outcomes, since, under this logic and philosophically speaking, it merely focuses on the power relations of North-North worldviews as part of a given (re) produced social structure, overlooking Global South ontologies. My research data provides such ontologies by addressing and interweaving stakeholders' narratives in light of 'Westernised' and 'native' mindsets, and my analysis and conclusions involve a local epistemology.

Post-structuralism centres on the discursive patterns to the production of subjectivity and identity, emphasising method and epistemological matters. This involves rethinking the conception of subject, which power and discipline produce (Alvesson and Sköldberg 2000). On the other hand, the transmodern approach frames my analysis, which centres on the possibilities of levelling power relations by considering ontological differentiations between Western and non-Western stakeholders, whom are part of the same phenomenon.

Even though the post-structural approach considers the interpretation of power relations, it does it under 'Western eyes' as reproducing coloniality of power/knowledge when addressing Global South study cases (as the post-colonial approach does) (see Grosfoguel 2011). This limits the acknowledgement of 'other' global worldviews (including 'Westernised' worldviews) in the Global South, which, in this case study, would as well limit the possible social benefits of the creation and implementation of local transport rules and policy. The transmodern approach thereby entails the opposite effect.

Key stakeholders' polarised social positions and power relations via interviews

I planned to interview key chosen stakeholders for my research in order to know their interests in improving the MCS. I carried out four interviews with a MCS Leader, a Public Servant from Tláhuac's municipality, an NGO Servant and a Scholar. I gathered key statements that reflect each stakeholder's opinions and knowledge about the importance of the MCS, which I translated as structural stories.

Defining key stakeholders' social positions

It is important to take into account that before I chose the four interviewees who participated in my research, some Public Servants I invited to take part in my interviews declined to participate. This is a starting point for my understanding of every participant's social position, as that conception can help me frame and visualise their motivations and needs in relation to the MCS. Moreover, I can analyse social positions of people who ended up not participating, since it can reflect individual interests other than those of the ones who participated in my interviews from the social 'institutions' they belong to.

In 2016, I invited the head of the planning area of the Mobility Ministry of Mexico City (SEMOVI) to be interviewed by e-mail. She showed interest in participating. However, later on, she declined. I, too, sent two invitation e-mails to the head of the development area of the Ministry and after handing an invitation letter to his office, I never received any answer from him. The case of the Public Servant from Tláhuac's Municipality was different from SEMOVI's Public Servants. It was complicated to get in contact with him but after I had delivered the invitation letter to his boss' secretary in person and called his boss directly, he agreed to the interview.

The difficulties involved in getting interviews from both Public Servants from SEMOVI can be seen as a sign of their distance from my study case phenomenon in a social position perspective. When each agent talks or acts from their occupied social space, they make use of their social capital from their social position (Larsen and Morrow 2009). Thereby, agents' contextual experience defines their everyday praxes from their social positions. These agents also tend to take reality/truth for granted by internalising praxes via institutionalisation of external structures (Freudendal-Pedersen 2009). Every social position could even entail a specific worldview on and for specific situations, such as my study case phenomenon. The central government of Mexico City is an institution in those terms that it could have different ideas and praxes about the city and the issues in its different zones.

The central and Western municipalities of the city tend to receive more central-governmental attention and public-private investment than the Southern and Eastern ones (see Finck Carrales 2015). That relates to the social position of the public servants who have the power to make it happen. Political and social interests and worldviews involve actions by public servants, for instance, dismissing or ignoring a public issue.

In the case of my MCS, the issue is about transportation in a power relational way. SEMOVI is apparently not interested in the MCS phenomenon of the peripheral zones because it is probably not part of their social position and worldview in terms of their political agenda. This is a conclusion that I made from the Public Servants' non-participation in my study together with some of the statements of the Public Servant from Tláhuac's Municipality and the MCS Leader interviewed, which I will mention later on.

Therefore, by not participating in the interviews, even though they were informed well in advance and it was made easy for them to participate, the public servants' decisions make me consider this non-participation to have political-social reasons. That means that the peripheral MCS phenomenon is 'politicised'. Politics is constantly intervening in the public agenda.

With regard to the interview with the Public Servant from Tláhuac's Municipality, he became alert and showed uneasiness about being interviewed perhaps due to political reasons. Nevertheless, his social position and worldview seemed very different from that of SEMOVI's Public Servants because he was very close to the MCSs of the municipality: 'As a user, well I use it very little, when I happen to use it, it is practically from one to two blocks and that's all [...]' [my translation] (Public Servant). Once in a while, he was an MC user himself. The Public Servant's social position physically relates to the MCS phenomenon. He has direct contact with practically every MCS organisation (MCSO) in the Municipality. As a member of the local government, he knows many dimensions of the phenomenon and that aspect has shaped his knowledge about the MCS drivers. This characteristic indicates that his social position makes him give high priority to the MCS phenomenon on his political agenda.

The case of the MCS Leader from the biggest MCSO in my study zone was very particular. After accepting right away to be interviewed, she gave me the most extended interview of all, which showed her clear interest in having it. Initially, she indirectly explained her social position from a historical starting point with regard to the MCS phenomenon: '[...] it [the MCS] starts to appear out of necessity, the necessity of having a job, the necessity of having an open job [...]' [my translation] (MCS Leader). I interpret this statement as a structural story: 'The MCS starts out of the necessity of people having a job'. She found herself in a position of social exclusion related to where she was born and had lived all or most of her life. The lack of employment opportunities at the peripheral zones of Mexico City has entailed that the inhabitants create their own ways of making a living from scratch (see *ibid.*). This characteristic probably creates a sense of belonging and responsibility towards the MCS of the study zone: '[...] for me it is a matter of morals that my work, which started twenty years ago, will one day culminate in a concession or permission from the Mobility Ministry where the work we do is recognized

[…]' [my translation] (MCS Leader). From her point of view there exists a distance between SEMOVI and the MCSs, which she has shortened over the years by demanding and asking for public attention. She knows that the only way of getting SEMOVI's attention and defending the project is through legal means, which could provide the MCSs with formality.

According to her, she has unveiled supposed threats and attempts of closing down and taking advantage of the service by the central and municipal governments on different occasions, and she used the help of lawyers in order to be able to continue with her work. In this regard, the distance between her MCS and the central government becomes more visible when she states the following: '[…] the Mobility Ministry […] has opted to deny us. We do not exist' [my translation] (MCS Leader). Her social exclusion goes beyond the individual, meaning that the ongoing exclusion reaches even people's working organizations. Apparently, the government has historically, directly or indirectly, created a social exclusion cycle that goes from social groups to their economic activities. In this case, this exclusion reflects in the legal frameworks of the city.

According to the MCS Leader, some Public Servants have taken advantage of the service's illegality by searching for patronised relationships with some of its leaders: '[…] to some groups […] it was said as an encouragement: "When I am there, you will be able to keep working and I will work for you to stay…" […] because new elections will come […] they have done this to us for twenty years' [my translation] (MCS Leader). She stated that the current 'disorder' of the service is the result of the constant free entrance of some MCSOs into the sector that do not take proper responsibility for and care of their activities. Additionally, some people from different central and local public institutions have allowed it in exchange for political support. According to the statement above, the government goes in and out of the MCS sector depending on their own political interests.

When the government is not regulating the service but at the same time gives hope for the possible regulation and legal permissions for working when election times come, causes the MCSOs to 'freeze' or 'paralyzes' them and makes them stay/stuck exactly in the social position in which they have been for the last twenty years as an informal working sector. This happens due to the government having a monopoly on legality creation (power) which reflects domination over the organisations in an historical way (see *ibid.*; Adey 2010).

In the case of the NGO Servant, the Latin-American Regional Head of the Institute for Transportation and Development Policy (ITDP), his social position seemed to be similar to SEMOVI's Public Servants with the difference that he accepted to be interviewed in his office located in a central neighbourhood of Mexico City.

> Well look, I'm not a specialist; I know that it is a service, which has answered to, well, a demand for mobility especially in the Southern-Eastern zone

[...] I don't know... I mean, I haven't gone to the zone, I've been told about it [...] It is a zone that, well... there is a lack of infrastructure [...] also, there have been many irregular settlements, etc.; it has a very particular topography [...] Let's see if, well! I'd like to go with you one day to see it with my own eyes, but, well yes, I think it's an important issue [my translation].

(NGO Servant)

His knowledge about the MCS of the city relates to positivist technical aspects of the study zone even though he has not been there in person. His physical distance from the phenomenon makes him understand it as something technical in most of its dimensions. That is something that is made up from 'scientific' *structural stories* about the Southern-Eastern zone of the city that could come from Western academia, the central government of the city, his NGO, other NGOs, private companies, etc. I can phrase them as follows: 'the MCS answers to a demand for mobility in the South–Eastern zone' and 'the South–Eastern zone lacks infrastructure and it has irregular settlements and a particular topography'. '[...] I have the impression that this [the MCS] gives you much more accessibility but it has maybe little capacity and it worries me also in terms of security [...]' [my translation] (NGO Servant). This comment reflects structural stories about the accessibility, capacity, and security of the MCS. He probably has those three concepts very well defined in his mind (whatever they are) from his institutionalised social position and worldview (structural stories: 'The MCS gives people much more accessibility' and 'MCs are dangerous'). This aspect indicates and stresses his ontological distance from the phenomenon because his relational thoughts are based on 'distorted' information that has been 'traveling' as discourses/narratives until reaching him.

Therefore, the NGO Servant's discourses/narratives about the peripheral MCS relate to Freudendal-Pedersen's (2018) explanation of people's preconceptions of other people's macro–political characteristics that make them the first to dismiss other people's structural stories that might be common. In this case, it is through geographical and social class separation that the structural stories about the service are prejudged in a power-relational way based on a 'Westernised mindset'.

In the case of the interview with the Scholar from the National Autonomous University of Mexico (UNAM), it is important to take into account that in 2014, he had carried out a positivist technical diagnosis of the MCS in the neighbourhood next to my study zone, so he already had considerable (mostly technical) knowledge about the phenomenon. His projects commonly tend to be coordinated with some central government ministries of Mexico City and external NGOs or private companies, so he knows how other key stakeholders think and act with respect to transport issues in the city. He works on behalf of his academic laboratory of Mobility and Green Infrastructure at UNAM, where he has a specific vision and aims:

[…] what we want is for the investment in the city to be more efficient […] that the investments in those nodes [peripheral] could be associated to this type of mobility […] the best way of creating proximity is through multidisciplinary projects and for these to have a more holistic vision in which mobility is one of the topics, not all the topics [my translation].

(Scholar)

His statement above involves his acknowledgement of a lack of 'access to the city' for inhabitants from peripheral zones (see Sheller 2014). By being aware of those technical aspects based on fieldwork research, I understand that these are different from the technical aspects of the NGO Servant from ITDP. I argue that is possible, since having *embodied* experience (body and mind affected by materialities [Jensen and Lanng 2017]) in the phenomenon creates different worldviews about it in people's minds. Such worldviews are different from the ones obtained just by being told about them as *'voice-to-voice' structural stories,* as they have been experienced rather than only heard. 'Well, I don't use it [the MCS], I know very little' [my translation] (Scholar). Additionally, the Scholar acknowledged himself as someone who knows 'very little' about the phenomenon based on the fact that he has not used the service. His social position and worldview, however, was modified at an ontological level from the moment he did his fieldwork research about the peripheral MCSs because it made him extend his understanding of the phenomenon through his own experienced "embodied" *structural stories.*

The possible regulation of the MCS: towards balancing power between social groups

Regarding the regulation possibilities of the MCS, each stakeholder's social position could explain their points of view as 'voices' in relation to possible future scenarios. For example, the Public Servant from Tláhuac's Municipality talked about the different interests of the MCSOs:

[…] it is difficult to deal with the ideas of every organizer, everyone has his/her own point of view and way… we don't intervene in how they lead their organizations, but we try to plan a good project in order for everybody to work in harmony with the municipality […] [my translation].

(Public Servant)

At the end, according to him, the 35 MCSOs of the municipality just want to provide the service after communicating their demands to the government of the Municipality. According to him, he has listened to the organisers' demands and has worked with them to the extent that his power and position will allow. With that, he has a main aim in mind:

The ideas are the dialog with people, with the organizers that, at the same time, take good care of the public […] that is one of the priorities, the good care; and that way, you can accept them in the future without any problem. [...] for trying to make a good project that benefits the community and, if in the future it reaches paperwork processes, to be part of it [...] [my translation].

(Public Servant)

His aim is that the service takes proper care of the users and, with that, to keep incorporating MCSOs. Even though he wants to develop the coordination of the services by the Municipality, the Municipality does not have the legal facilities for doing it. Nevertheless, according to him, he tends to listen to the demands of MCSOs leaders and tries to make agreements with them towards a more likeable and safer service for the public (users and non-users).

In this regard, the MCS Leader has implemented and abided by the MCSs 'informal regulations' that have come from agreements between the Municipality and the organisations as something positive:

'Here in Tláhuac, a regulation was developed 3 years ago [...] with it, we told to the youth [MC drivers]: "This is what the authority dictates and we have to follow the rules". It worked very well for us [...]' [my translation] (MCS Leader). Seemingly, the regulations were actually partly a coordinated effort that came from the willingness of the Municipality and the organizations. Her statement indicates that having the rules have helped them to have certain control and coordination over the different MCSs in the study zone, especially over their drivers (structural story: 'Rules help with the coordination and control of the MCSs'). The creation of rules and its implementation coming from locals, therefore, are necessary for the improvement of the service, since these seem to balance the power distribution between social groups as, in this example, the MCSOs and the government have agreed on common rules.

'Today, we know that according to the Law, [...] the permission should be given by SEMOVI [...] the disordered way is what causes troubles' [my translation] (MCS Leader). I interpret a structural story from this statement: 'The disordered way of providing the MCS involves troubles'. Regardless of the conflicts between the government of the Municipality and its MCSOs, she somehow and to some extent has had a 'voice' to be listened to when attempts of regulation have been made. Moreover, she has participated together with other MCSO leaders in several attempts of what could be the starting of service formalisation in the City Congress.

[...] four years ago, more or less, the intervention in the [Mobility] Law that gave us at that time the President of the Mobility Commission of the Congress of Mexico City. Together, we had several meetings, many I would say, in order to agree on this new Law that would make it possible to start our regulation [my translation].

(MCS Leader)

Some of the MCS leaders in the study zone have met with Congress members who have been in charge of the transport regulations of the city. Nevertheless, the MCSOs have had to learn how to defend themselves legally when the central government of the city (the executive branch) put pressure on them with regard to their illegality. She also acknowledges that in order to create the regulations, it is necessary to have knowledge about the phenomenon, such as formal diagnoses and studies. 'There have been studies where [...] they [the government] tell us and everything, but we don't know the results and they haven't told us [...]' [my translation] (MCS Leader). The central government has told the organisation about apparent studies on MCSs that it has carried out without inviting the organisation to take part of or even sharing their results with them. This suggests a clear separation between each social group, even when the leaders have been invited to meetings in the City Congress.

> I would like one day for the Government to let us express ourselves like that. They always give us 5-10 minutes and they tell us and good-bye, ready, bye. No! There should be open tables where we could really express ourselves. I know that many of us don't have that skill because [...] we haven't got the required education, and the Government doesn't let us express ourselves that way, but believe me, we have the knowledge and willingness [my translation].
>
> (MCS Leader)

The quote above is divided in two interesting statements that are interrelated to social position, involving an epistemological issue: on the one hand, when the MCSOs have gone to the City Congress, they have had little 'voice' in the meetings. According to the MCS Leader, they have been invited to take part in the meetings, but they have not been completely listened to by the government, which probably makes the government dismiss the needs, issues, opinions, ideas, and even knowledge of the MCS leaders about the MCS phenomenon (structural story: 'The City Congress has not listened to the MCS leaders'). On the other hand, the MCS Leader considers herself as less educated than the Public Servants, especially the Congress members. Nevertheless, through my studies and the interview outcomes, the leaders have proved to have much more information and knowledge (mostly empirical) about the MCS phenomenon than the public servants of the city I have been in contact with.

In this case, the difference of social position between the social groups is mainly based on their socioeconomic and political power levels, which, at the same time, are based on their geographical position and social class within the city (structural story: 'people's power level in mobility and transport issues for its regulation depend greatly on their individual social positions').

In the case of the NGO Servant from ITDP, his social position could apparently be between the ones of the MCSs leaders and the public servants, since their institution has a certain extent of power regarding transport issues.

[...] if it could be possible, we as an organization have always supported very much [...] the topic of the three-wheel freight and also for two-person carriage. [...] in other cities, in ITDP-India, it helped improve the design of the rickshaws [...] it was a great victory that these weren't banned [...] [my translation].

(NGO Servant)

The NGO Servant explained that his institution has gotten involved in other countries' cases of informal transportation, where they had carried out studies and even political interventions about it. However, for some reason, ITDP-Mexico has not been interested in intervening in the MCS of Mexico City even though ITDP is very close to SEMOVI and the City Congress. The NGO Servant's social position therefore is closer to the central government. The relations between ITDP and SEMOVI and the lack of relations between ITDP and the MCSOs are core points within power distribution for service regulation possibilities.

In this regard, the Scholar from UNAM had an interesting statement about why the central government of the city had not gotten close to and, consequently, had not regulated the peripheral MCS yet:

[...] there is a social class barrier, mainly towards the decision makers on the governmental level because on the one hand you have that they want to 'do' as they 'do' in Denmark, and Denmark is usually one 'mental regression' and one reference of the decision makers, but they don't realize that bicycle mobility is seen in parts of [Mexico] City's East where there are very few resources and where it is a lifestyle, and when they have the opportunity they 'beat' them [the MCSOs]. [...] [A]s a country, when we talk about mobility, our decision makers have an emotional dependency [...] it really affected them to have been in Denmark [...] they told us: "No, it should be as in Denmark", and we, well: "It cannot be as in Denmark because this is Mexico". But there are still many decision makers there in the key sectors, who keep thinking that it has to be like in Denmark without taking into account that the problem of social justice in Mexico happens before, in an urban way, in the access to housing and dignified work. [...] it is a problem of social class, I'm telling you, regarding the professions/trades. Thereby, currently an MC driver and an UBER driver cannot be seen the same way, but actually they are the same, they are providing the same service. [...] I think that in order to improve, the service would need a legal status, that is the main thing, [...] recognized by SEMOVI. They [the public servants] should stop looking down on them [the MCS workers]. I imagine that if they were like those blonde guys who are there, in 'Strogetts' [a European street] waiting for people, they would see them in a different way. [...] it is very odd because the same public servants, who go there [Europe] and get in and use them [the bicycle cabs], are the same who are 'beating' the [MCSs] groups here. Therefore, SEMOVI in that sense has a great challenge [my translation].

(Scholar)

This statement addresses different dimensions as a transmodern self-valuation aspect that could explain the lack of MCS regulations in the city. The basis is the social position from the acknowledged differentiation of social classes and skin colours. However, his explanation goes many steps back by tackling ontological aspects connected to Global North-Global South relations: as explained before, transport studies and projects in Latin-America tend to be adaptations of other European/Anglo-American ones. In my study case, for some reason and according to the Scholar from UNAM, as a structural story, 'transport public servants of Mexico City identify more with European cities'. They want to force-adapt what they see in Europe into Mexico City, and apparently, the issues that are not similar to that of Europe are completely invisible for them, such as the peripheral MCS. What they see in Europe translates into 'voice-to-voice' structural stories from a 'Westernised mindset': some Public Servants go to Europe; when they go back to Mexico, they relate how well things are done there to other Public Servants, scholars, entrepreneurs, etc., from their own social position creating solid stories in other's minds about it, in this case, transport policies.

The possibilities of the MCS regulation therefore part from the ontological level of acknowledging, in the first place, the existence and importance of the phenomenon in contextual terms and, then, consequently opening the possibilities for giving 'voice' to the people who are part of it, such as leaders, drivers, and users (a pluriversal opportunity). That could be done in order to balance the power of the phenomenon.

Conclusions

Methodological and conceptual contextual outcomes

The non-participation of governmental representatives in my interviews involves an ontological issue regarding the formalisation of MCSs. Public servants of the central government of Mexico City tend to have a social position related to high-class society in social networks and geographical terms. That social position relates and gets close to European/Anglo-American worldviews regarding transport planning, and the public servants prefer to look at how this is done in Europe instead of looking at the worldviews of their own culture. This involves them having a 'Westernised mindset'. When social problems are outside those social positions, these do not exist for people that form those social groups or, if they know them, they dismiss them or if possible try to make them disappear.

Inspired by the studies of Larsen and Morrow (2009) and Bourdieu (1996), the problem I draw upon is not what people's social positions are but what people's actions based on such social positions can potentially involve and favour in relation to a certain issue. In other words, it is important to look at what an individual can do about an issue from his/her social position besides defining or being aware of those positions. Most importantly, the transmodern approach involves considering power structures and relations within those social positions in order not to reproduce knowledge of the Global North and to favour additional outcomes in the Global South that can impact societies in a more 'positive' way. By incorporating locals' worldviews about the phenomenon as a pluriversal opportunity, this entailed a Global North-Global South dialogue.

Furthermore, based on Freudendal-Pedersen's (2009; 2018) research, I differentiate between three types of structural stories: embodied (individual body and mind experience), voice-to-voice (stories travelling and changing between social groups as word of mouth), and scientific (outcomes of studies with good reputation). The importance of these categories rests on structural stories shaping people's perspectives and opinions about a phenomenon towards their worldviews, so these are significant when relating them to contextual social praxes at an ontological level. Those categories make it easier to understand their sources, possible social relations, power displays and implications, and even consequences of the phenomenon to which they belong to facilitate the creation and understanding of a pluriversal 'dialogue'.

Recommendations for public policy and legal framework of the MCS

One 'barrier' for reaching formalisation in the MCS is that being geographically placed in the peripheries of Mexico City, the neighbourhood undergoes urban and social fragmentations, which forms ontological separation between social groups. This situation has entailed ineffective or no action by the government in relation to the formalisation of the MCS.

Therefore, while planning based on positivist approaches suggest that transport projects should adapt to guidelines and propositions manufactured in the Global North, I argue that it is important that governmental representatives personally go to the neighbourhood, use the MCS themselves and afterwards take part in further multi-social groups' meetings to reshape their worldviews about the phenomenon. Governments have to make sure to listen to the statements provided by the MCSOs beyond fulfilling public meeting quotas. The different social groups have to meet physically and have to listen carefully to one another in the form of a pluriversal 'dialogue'. Here, the power relations could balance from an ontological acknowledgement to an active one, towards groups sharing the same epistemology on the phenomenon.

Tláhuac municipality has historically 'politicised' the MCS, but it has also been physically and ontologically closer to the phenomenon and has created local 'informal regulations' that have improved the service. In Mexico City, centralised mobility and transport policies and regulations have favoured control by the central government (executive and legislative branches) over excluded social groups, especially regarding their self-created economic activities. Giving more power (attributions) to the municipalities in this regard could terminate that political control.

The Constitution of Mexico City (Article 59) states that underserved social groups in the city (from 'native neighbourhoods') can have autonomy/'self-determination' in some of their activities that can involve 'economic, social, and cultural development'. Therefore, based on this, the law should give more autonomy to the MCSOs.

Additionally, the regulation should state punishments for those who try to patronise and intimidate workers of the service. This can prevent the disparities of political parties between the municipalities and SEMOVI in which the MCS has been taken as 'political hostage'.

In 2017, a formalisation process of the MCS started in Mexico City. By 2019, Public Servants of SEMOVI approached to me, asking for consultancy with regard to informal transportation. I shared with them the conclusions I highlighted in this chapter, which they are considering for the latest regulation of the MCS.

References

Adey, Peter. 2010. *Mobility*. New York: Routledge.

Alvesson, Mats, and Kaj Sköldberg. 2000. *Reflexive Methodology, New Vistas for Qualitative Research*. London: SAGA.

Ateljevic, Irena. 2013. "Transmodernity: Integrating Perspectives on Societal Evolution." *Futures* 47, 38–48.

Beaverstock, Jonathan V., Richard G. Smith, and Peter J. Taylor. 1999. "A Roster of World Cities." *Cities* 16, no. 6: 445–458.

Bourdieu, Pierre. 1996. *Physical Space, Social Space, and Habitus*. Rapport 10, Institutt for sosiologi og samfunnsgeografi. Universitet i Oslo.

Büscher, Monika, Mimi Sheller, and David Tyfield (eds.). 2016. "Mobility Intersections: Social Research, Social Futures." *Mobilities* 11, no. 4: 485–497.

Dussel, Enrique. 2012. "Transmodernity and Interculturality: An Interpretation from the Perspective of Philosophy of Liberation." *Transmodernity: Journal of Peripheral Cultural Production of the Luso-Hispanic World* 1, no. 3: 28–59.

Dussel, Enrique. 2016. *Filosofías del sur. Descolonización y transmodernidad*. México: Akal/ Inter Pares.

Escobar, Arturo. 2016. "Thinking-Feeling with the Earth: Territorial Struggles and the Ontological Dimension of the Epistemologies of the South." *Revista de antropología iberoamericana* 11, no. 1: 11–32.

Farrés Delgado, Yasser and Alberto Matarán Ruiz. 2014. "Hacia una teoría urbana transmoderna y decolonial: una introducción." *Polis*, Revista Latinoamericana, 13, no. 37: 339–361.

Finck Carrales, Juan C. 2015. "Transporte Emergente en colonias periféricas de la Ciudad de México: El caso de los Bici y Moto-Taxis en La Conchita Zapotitlán, Tláhuac." Tesis de Maestría en Proyectos para el Desarrollo Urbano, Universidad Iberoamericana, Ciudad de México.

Freudendal-Pedersen, Malene. 2009. *Mobility in Daily Life: Between Freedom and Unfreedom.* Farnham: Ashgate Publishing, Ltd.

Freudendal-Pedersen, Malene. 2018. "Engaging with Sustainable Urban Mobilities in Western Europe: Urban Utopias seen through Cycling in Copenhagen." In *Handbook of Anthropology and the City*, edited by Setha Low, 240–253. London: Routledge.

Galland, Daniel and Pablo Elinbaum. 2018. "A 'Field' Under Construction: The State of Planning in Latin America and the Southern Turn in Planning". *disP—The Planning Review* 54, no. 1: 18–24.

Gehl Architects. 2009. *Transforming Mexico City into a cycling city.* http://gehlpeople.com/work/cases/.

Grosfoguel, Ramón. 2011. "Decolonizing Post-Colonial Studies and Paradigms of Political-Economy: Transmodernity, Decolonial Thinking, and Global Coloniality." *Transmodernity: Journal of Peripheral Cultural Production of the Luso-Hispanic World* 1, no. 1.

Haraway, Donna J. 1991. *Simians, Cyborgs, and Women. The Reinveintion of Nature.* London: Free Association Books.

Jensen, Ole B. and Ditte B. Lanng. 2017. *Mobilities Design: Urban Designs for Mobile Situations.* London: Routledge.

Larsen, Kristian, and Virginia Morrow. 2009. "Social position and young people's health: A Bourdieu'ian critique of dominant conceptualisations of social capital." *Praktiske Grunde* 9, no. 3: 7–30.

Mignolo, Walter D. and Catherine E. Walsh. 2018. *On Decoloniality: Concepts, Analytics, Praxis.* Durham: Duke University Press.

O'Donnell, Loretta, Robin Kramar and Maria Dyball. 2013. "Complementing a positivist approach to investment analysis with critical realism." *Qualitative Research in Financial Markets* 5, no. 1: 6–25.

Porter, Libby. 2006. "Planning in (Post) Colonial Settings: Challenges for Theory and Practice." In *Planning Theory & Practice.* Routledge.

Quijano, Aníbal. 1991. "Colonialidad y Modernidad/racionalidad." *Perú Indígena* 29, 11–21.

Roy, Ananya. 2009a. "Strangely Familiar: Planning and the Worlds of Insurgence and Informality." *Planning Theory* SAGE Publications, 8, no. 1: 7–11. DOI: 10.1177/1473095208099294.

Roy, Ananya. 2009b. "The 21st-century Metropolis: New Geographies of Theory." *Regional Studies*, Routledge, 43, no. 6: 819–830. DOI: 10.1080/0034340070 1809665.

Sassen, Saskia. 2005. "The Global City: introducing a Concept." *The Brown Journal of World*, Affair winter/spring, XI, no. 2: 27–43.

Sheller, Mimi. 2014. "The New Mobilities Paradigm for a Live Sociology." *Current Sociology* 62, no. 6: 789–811.

Solares, Blanca. 2009. "La Diosa en la cosmogonía mexica." In *Aportaciones al Estudio de la Cosmovisión*, edited by Fernando, Sançén Contreras, Ciudad de México: UAM-X.

Tlostanova, Madina. 2017. "On Decolonizing Design." *Design Philosophy Papers* 15, no. 1: 51–61.

Tlostanova, Madina, Suruchi Thapar-Björkert and Ina Knobblock. 2019. "Do We Need Decolonial Feminism in Sweden?" *NORA—Nordic Journal of Feminist and Gender Research* 27, no. 4: 290–295. DOI: 10.1080/08038740.2019.1641552.

Watson, Vanessa. 2016. "Shifting Approaches to Planning Theory: Global North and South." *Urban Planning* (ISSN: 2183–7635), 1, no. 4: 32–41. DOI: 10.17645/up.v1i4.727.

Yiftachel, Oren. 2006. "Re-engaging Planning Theory. Towards 'South-Eastern' Perspectives." *Planning Theory*, SAGE Publications, 5, no. 3: 211–222. DOI: 10.1177/1473095206068627.

Yiftachel, Oren. 2016. "Extending Ethnocracy: Reflections and Suggestions." *Cosmopolitan Civil Societies Journal* 8, no. 3: 52–72. http://dx.doi.org/10.5130/ccs.v8i3.5272.

3 Decolonising global health promotion

A quest for equity

*Rashmi Singla, Johanne Andersen Elbek
and Lene Maj Hjortsø Fernando*

Introduction

A consistent central element of global health is the matter of equity; a discussion with roots in colonialist endeavours, equally relevant in all levels of the field today and in the future. Even with just intentions, unjust structures are still a reality in the foundation of global health at present. For global health to be based on principles of equity, the structure of global society ought to be based on equity too. Only this enables a fully integrated implementation of global policies that seek to promote health and equity for all. With current Eurocentric societal structures enabling a continued exclusion and suppression of the Global South, this chapter presents the need to decolonise the field of global health in the quest for health equity and as such highlights *four essential levels in the decolonisation process: knowledge, leadership, policy, and praxis.* With this chapter we attempt to move across time and space, and in doing so appeal for collaborations and point to barriers and transformations on multiple levels through an array of empirically based studies from both the Global South and North. Adding to the emerging debate on how to approach the decolonisation of the global health field, we argue that it may be time for scholars and professionals to move away from universalism and towards pluriversalism, and by doing so, acknowledge the multiple ways of being and doing that have always existed. Through challenging the implicit notion of a 'one–size–fits–all approach', we confront power differentials and prevent reproducing the universalism that is inherent to coloniality. Throughout the chapter we aim at illustrating how true decolonisation in the realm of global health may require a foundation reset of the global health concept and that its content and premises be reimagined. We suggest that the future of global health must strive for decolonising changes on all levels, with a focus on ethical considerations and pragmatism—a future in which the Global South position and understanding of knowledges, health, and health promotion practices are equally valid.

First author, Rashmi Singla, holds a doctorate in Psychology from the Global North, Denmark, and originates from the Global South, more specifically India, from where she received her first master's degree. Yet having

DOI: 10.4324/9781003172413-4

lived for 41 years in Denmark, she has a dynamic sense of belonging to both contexts, moving beyond the conception of 'thinker thinking from the South' (Grosfoguel 2011, p. 3). The two co-authors are MSc students within the field of health promotion and international development who originate from the Global North, Denmark. Common to the three of us is that we sympathise with a pluriversal perspective that supports global equity, and we collaborate on this chapter with inspiration from the course at Roskilde University, Denmark delineated later in the chapter.

The chapter begins by explaining differing understandings and interpretations of the ever-changing concept of global health that is closely tied to global socioeconomic development. We then present relevant global health policies and discuss power structures in the policy and knowledge foundation of the field with emphasis on global equity being the goal. Throughout the chapter, several challenges in achieving this goal are outlined, including the dominating paradigm of neoliberalism that supports a capitalist orientation of development while neocolonialist tendencies in all levels of the field maintain a structural suppression of Global South voices. Through historical consciousness of colonial legacies, the chapter highlights the roots of contemporary global inequity in the field. The chapter proceeds to discuss current dynamics related to human encounters in promoting health globally, foregrounding mobile health professionals, specifically challenging the conception expatriate. Finally, the chapter reflects on concrete practical approaches and presents broad recommendations for future work in the realm of global health.

The concept of global health

In a globalised world where nations are increasingly interconnected as one global community, the field of global health is more central than ever. Global health as a field of study has existed since the late 1970s and has since increasingly gained more and more recognition as an academic discipline (Stewart and Swain 2016). While scholars define the concept in various ways, we support a broad definition, acknowledging that there are multiple aspects to global health. Global health is tied to emerging global shifts in health status and convergences across the globe. In addition, global health is also defined by technical understandings, in terms of the control and prevention of particular diseases, including communicable and noncommunicable diseases. As a generally accepted underlying premise, global health strives to be a transnational cooperative focusing on promoting health for all (McCracken and Phillips 2012).

The field of global health is constantly evolving in close association with globalisation. The term global health has gained grounds over public and international health, alongside a greater awareness that the health and well-being of people across the globe is increasingly interrelated and interdependent (*ibid.*). Within the temporary scene of global health, health has therefore also become a commodity in the global marketplace (Dutta 2008).

The most current example is the global handling of the Covid-19 pandemic, where pharmaceutical organisations and nations both cooperate and compete against each other, in order to produce and ensure the purchase of an effective vaccine (Bochkov 2021). Additionally, people's health is a precondition for physically, mentally, and socially well-functioning employees, referring to the WHO definition of health as 'a state of complete physical, mental and social well-being and not merely the absence of disease or infirmity' (WHO n.d). Hereby, the health of people is essential for the functioning of the globalised societal structure we live in. This fact has only become clearer during the Covid-19 pandemic; the societal impact of the pandemic has been and most likely will continue to be serious in relation to both health care systems, economy, and health of global citizens. As brought forth by the Danish Prime Minister during the pandemic: 'Health and economy are not each other's opposites – they are each other's prerequisites' (Frederiksen 2020). While the total consequences of the pandemic are yet to be examined, one thing is certain—it has highlighted the importance of global health. A striking example of this landmark change is The Great Reset initiative launched in 2020 by the World Economic Forum. They articulate a need to rethink capitalism and '...what we mean by "capital" in its many iterations, whether financial, environmental, social, or human' (Schwab 2020). The Great Reset is a post Covid-19 paradigm that entails pragmatic steps towards building a more resilient, cohesive, and sustainable world, where all stakeholders, including businesses, governments, and civil society groups, must work for shared values of equality, health, and environment to combat global challenges and risks (*ibid.*).

Already in the 1990s scholarly work has raised awareness about the health and inequity consequences of globalisation (Navarro 1999) and today various scholars and studies support the idea of and need for a paradigm alteration towards a more sustainable development approach with emphasis on structures and solutions that enhance equity, environment and the health of people and the planet (Baum 2009; Sunyer and Grimalt 2006; WHO 2009). Such a societal structure stands in stark contrast to the current neoliberal order, where transnational corporations are the main beneficiaries of globalisation and continue to play a major role in the realm of global health. The transnational hegemonic configuration influences the global agenda with interests of elite national actors and profit-driven organisations (Dutta 2008). Yet movements are surfacing in both macro, meso, and micro levels, with examples from research studies, more integrated sustainability and health promoting policies and regulations, consumer awareness and activist groups. With evident climate changes increasingly dominating the global agenda, the field of global health is changing its focus, incorporating planetary health into policies and recommendations (Sunyer and Grimalt 2006). Further, the contemporary framework of global health has continuously grown to become a more holistic concept with health promotion as a central element. It is now widely recognised that the field of global health not only seeks to oversee and manage global diseases and physical health in the biomedical sense, but also engages in the way we live, work and interact with each other in a globalised world (Dixey, Warwick-Booth, and South 2013).

It is our belief that in working with global health, equity ought to be the fundamental principle. With equity we mean both health equity, giving all human beings a fair and just opportunity to be healthy, and equity in the involvement in both knowledge production, leadership, policy, and praxis. Global equity is a prerequisite for equity in global health. Ensuring equity is also to acknowledge pluriversalism, allowing a plurality of ideologies to co-exist with no dominating universalistic point of view. This is in line with Ashish Kothari et al. (2019) who define a pluriverse as 'a world where many worlds fit'. Although global health considers the entire human population as one global community, it continues to differentiate between Global South and North, which in this chapter are referred to as divisions of the world nations understood as neither spatial nor geographical, but rather as hierarchical relations of power between North and South (Richey 2015).

From the above range of interpretations, we want to convey the understanding that global health must advocate health and equity for all people worldwide, while the concept of global health is rooted in historical development, interlinked and susceptible to many global development concerns, both in regard to politics, climate, economy, security, and beyond.

The colonial legacies of global health

The very outset of this book is that a narrative adjusted to dominant colonial interests, having little to do with the diverse social realities and histories in which they have emerged, sustains dominant approaches to global problems, and this affects the possibility of transcending these very same problems. A focus on the history and current social dynamics is necessary to rethink the relationship between past and present in contemporary global health policies and encounters for scholars and professionals. While the attention to the concept of global health is growing more than ever, recent criticism points to the fact that in the majority of academic institutions around the world, the discipline of 'global health' is taught in predominantly depoliticised, uncritical, and ahistorical ways (Saha, Kavattur, and Goheer 2019). While the global health institutions of today work for an equal and just world, the focus and prevalence of the discipline has developed with little critical reflection of the historical legacies on which the concept of global health is based. The need to decolonise global health, we argue, is the need to break with the unconscious dependence on models of thinking, making, and interpreting the world 'on the norms, created and imposed by/in Western modernity' (Tlostanova, Thapar-Björkert, and Knobblock 2019, 290). In the encounter between humans this also implies to break with 'racism, hetero-patriarchy, economic exploitation, and discrimination of non-European knowledge systems' (*ibid.*).

The phenomenon of global health practice as we know it today, developed from roots stretching back centuries into the past and pioneers of global health can be traced directly back to European colonialists. Through critically examining the empirical power and politics of colonial medicine (Greene et al. 2013), it becomes clearer how both contemporary health inequalities and mistrust in Western health interventions came to be. In the history of colonialism, medicine functioned as a dividing force with the superior Western practitioner on the one side, and the savage and diseased nonwestern patient on the other. The Global South in many cases served as a testing ground for the development of both medical and public health research practices. Observations of disparity in infectious disease mortality between coloniser and colonised served to legitimate imperial projects, and over time this developed into racial hierarchies based on embodied and biological characteristics (*ibid.*). It was through colonial medicine that the first international epidemiological investigations were initiated, allowing to draw comparisons between health and disease across various continents (*ibid.*). This can be argued to be the first step towards global health. Such international epidemiological comparisons led to the birth of the concept of 'tropical medicine'. While the concept engaged with geographical divergence and not social, it ultimately contributed to the establishment of a racial division between the white and black body, leading to marginalising suffering as an effect of culture, rather than disease or poverty (Keller 2006). The social and cultural framing of ill health very much persist to this day. With the birth of tropical medicine, we can begin to explain why the Global North automatically understands global health as challenges and acts taking place geographically distanced from 'us' and focuses on someone else; thus, leading to an 'othering' of 'them'. The international health scene following has been governed and financed through an alliance of powerful Western organisations for decades, thus reinforcing the colonial lens of North to South cooperation (*ibid.*).

Understanding contemporary global health is also understanding the roots of knowledge behind health and healing. Parallels of the past and present might be a result of divergent understandings of the nature of sickness and health, and ultimately rooted in divergent epistemologies between the Western practitioner and the colonised populations. In the colonies, Westerns medicine received official support, while local and traditional systems were consciously suppressed, creating clear divisions between legitimate and illegitimate directions for health development (Greene et al. 2013). With Western governance, alternative ways of understanding health and disease were simply rejected, and Western medicine deemed the only scientifically valid way. As such, there has been a worldwide domination by European forms of organisation and scientific systems throughout history (*ibid.*). 'Any Western medical institution more than a century old and which claims to stand for peace and justice has to confront a painful truth—that its success was built on the savage legacy of colonialism.' (Horton 2019, 996).

While the colonial legacies have burdened previous generations in the Global South, it has also led to an indispensable advancement in the development of global health. Needless to say, the formation of the World Health Organization (WHO) in 1948 has been an essential step in the development and future of global health. With a decentralised structure of governance, the WHO moved decision making closer to the users of health services around the world (WHO 2020). As an official advisor and coordinating body in global health, WHO has been the reason for major health advantages, including the successful global eradication of smallpox, which is still today known to be the great triumph of modern global health efforts. Despite the WHO's undeniable successes, critics point out a continuous colonial discourse as global campaigns were built on primarily Western interference in local populations, where in several cases they met strong resistance and compromised the agency of the targeted populations (Greene et al. 2013). Jeremy Green et al. (2013) have highlighted how violations of human rights within the interventions may have compromised the work of future global health efforts, and possibly left a residue of mistrust towards future interference. Colonial legacies have led to several global health interventions and programmes that, due to history, have met resistance in former colonised nations (*ibid.*). According to Richard C. Keller (2006, 1), today many patients in the Global South thus still associate Western health interventions and professionals with both political and economic power, while on the other side of the encounter the subjects of intervention are viewed as unruly, superstitious and in many cases as 'incapable of compliance with biomedical regimens'. Such suspicion of Western medicine intentions reinforces a reciprocal mistrust in between health professionals. Today, a Western discourse still exists, anticipating that Global South populations cannot fully manage Western medical practices and treatments, despite the fact that an increasing number of studies demonstrate the opposite (*ibid.*). Meanwhile the Southern nations' development paths have been disrupted, as the very same discourse gives birth to a sense of inferiority among the colonised. Hence, legacies of mistrust pose a great threat to both present and future global health efforts. Therefore, we must be salient about the parallels between contemporary global health and colonial medicine, and indeed how global health governance structures came to be in the first place. The fact that the mindset of colonial times has been carried forward into that of global health today is evident in the harsh truth that the global health agenda is still dominated by the Global North (Reidpath and Allotey 2019). Using a neocolonial lens, that is being aware of the continued indirect control of marginalised nations, the Western man is building careers on studying the health and illnesses of vulnerable groups from the Global South, which strengthens global racial hierarchies and the notion of othering. As such, social structures result in specific groups being systematically privileged and disproportionately able to control the generation of scientific knowledge and advancement within health and medicine.

In many ways, the foundation of global health and modern medicine grew out of globalisation of both science, commerce, and politics in the mid to late 19th century (Greene et al. 2013). Development and sharing of ideas, medical traditions, and knowledge was made possible by global interactions and assimilations carried out by colonising powers. Through the evolution and advancement of medical practices, drug use and theories of diseases colonising powers inevitably led to the making of modern Western medicine (Chakrabarti 2014). The legacies of colonial medicine led to growing differences, which can still be identified within health care, preventive medicine, and mortality rates between former colonies and colonisers; disparities in health status and social injustice found in the Global South today; evidently consequences of colonial times (*ibid.*). Although one can argue for the dominant discourse of global health today being altruistic, the origins of medicine and global health have rather been tools of diplomacy (*ibid.*). As the ways in which the European imperial endeavour reflected unequal power relations in colonial times, so too the unequal power relations within the practice of global health reflects the inequity of the present (Saha, Kavattur, and Goheer 2019).

The development of global health promotion

The practice of health promotion has developed accordingly with the changing concept of global health. The concept of health promotion traditionally set out to prevent disease and illness and thus had a treatment-oriented focus, and now encompasses various global policies that seek to create awareness and present recommendations on promoting health worldwide. With the 1986 Ottawa Charter, the first international agreement on regular health promotion was signed, introducing a focus on 'health in all policies'. The conference was built on the growing expectation for a new public health movement and primarily aimed at challenges in the more industrialised nations, prompting a need not just for prevention of illness, but also the promotion of good and healthy lives. According to WHO (2016), health promotion 'enables people to increase control over their own health. It covers a wide range of social and environmental interventions that are designed to benefit and protect individual people's health and quality of life by addressing and preventing the root causes of ill health, not just focusing on treatment and cure'.

With the evolving understanding of health promotion, contemporary global health policies seek to grasp all the different facets of healthy living, related to the context of the individual, community, and planetary health. Common to global health policies is that they have moved away from a focus on risk factors and instead aim at empowering people to make healthy choices and reducing health inequities (WHO 2009). The Millennium Development Goals launched in 2000 by the United Nations (UN) highlighted the health inequalities between the Global North and South, and prompted the need to mobilise resources for health in poorer countries (WHO 2005). In recognition of the global inequalities in health, a greater understanding of the underlying causes evolved. A growing body of research supporting the existence of the Social Determinants of Health meant a shift in focus from looking at how health affects economic status to quite the opposite; how economic status affects one's health. WHO establishing the Commission on Social Determinants of Health in 2005, meant that global health institutions officially acknowledged social status as being directly connected to health and well-being (Dixey, Warwick-Booth, and South 2013). With this also came a greater understanding of why global health inequalities continue to be the major challenge caused by an unequal world. 'Leave no one behind' is the promise of the agenda in the Sustainable Development Goals (SDGs) launched in 2015 by the UN. The holistic agenda was a call for renewed intersectoral action, recognising that global health also encompasses processes, policies and programmes outside the health sector that through social, economic, political, and environmental determinants all have implications for health (Hussain et al. 2020). Through the development of global policies, the understanding of health has evolved to comprehend every aspect of human's lives and the contexts in which we live.

Although the highlighted policies place health on the global agenda, there is no universal implementation strategy, nor should there be, as local settings and context is fundamental in designing health promoting interventions. There are naturally differences within and between countries in terms of social determinants (Marmot 2005) and intervention approaches, particularly between low- and high-income countries. Where national strategies in low-income countries may focus on basic needs for a life without disease, high-income countries may have more 'privileged' challenges that involve improving health rather than avoiding illness, as a result of a greater socioeconomic development. The common denominator of health promotion interventions is the focus on promoting and facilitating healthy, safe choices for all people, regardless of social status. Health promotion requires a stable foundation in several preconditions '...peace and social justice; nutritious food and clean water; education and decent housing; a useful role in society and an adequate income; conservation of resources and the protection of the ecosystem' (WHO 2009, 10).

On a local level, different intervention approaches seek to facilitate behavioural changes. As a consequence of the Western dominated development of global health, the most qualified temporary tools are predominantly founded in knowledge produced in the Global North. We choose to highlight the setting-approaches as an example of an overall intervention strategy, as we argue these are valuable approaches in ensuring local context relevance in both Global South and North settings. The setting-approach and its successor, the supersetting-approach, constitute the foundation for an integrated, local-based intervention. The setting-approach was founded in the 1980s based on WHO policies. It was used as a method to integrate health promotion in the settings of people's everyday life in both individual, social, and structural terms (Bloch et al. 2014). The supersetting approach adds on to the setting-approach with increased focus on longer term interventions and greater interaction and interconnectedness between the various settings in the local community. The supersetting approach leans on the Adelaide Declaration 'health in all policies', which proposes a new type of governance that integrates health promotion between all levels of government, across sectors, and partnerships in civil society and the private sector (*ibid.*). The values and principles of the approach '...are associated with the concepts of integration, participation, empowerment, context and knowledge' (*ibid.*, 10). Further, it may largely facilitate the currently limited knowledge-sharing of Global South communities, as the interventions take place in a co-creating process. This is an essential factor in global health promotion practices, because interventions need to be context specific in order to make a difference for the stakeholders involved (*ibid.*). As it is highlighted later in this chapter, this approach may be valuable for health professionals in the quest to include the perspective of the local, and ultimately it may also allow for an alteration of the terms that constitute the field of global health.

Contemporary challenges to health equity: social determinants, climate change, and Covid-19

The dominating neoliberal paradigm enables neocolonialist tendencies in all levels of the field of global health, sustaining inequity as a key global health challenge. In the quest for equity in health, we point to three examples of contemporary challenges: social determinants, climate change, and Covid-19.

According to WHO: 'Health equity is defined as the absence of unfair and avoidable or remediable differences in health among population groups defined socially, economically, demographically or geographically' (WHO n.d.). *Social determinants* can influence health equity positively or negatively and studies show that they may be even more influential than health care or lifestyle choices. Social determinants involve 'the conditions in which people are born, grow, work, live, and age, and the wider set of forces and systems shaping the conditions of daily life' (*ibid.*). Common to people from both the Global South and North, the social gradient applies to their health statuses: the lower the socioeconomic position, the worse the health (*ibid.*). The aforementioned context of globalisation, has led to the slowing down of health improvements that occurred worldwide since the last decades of the 20th century, causing stagnation and even decline in the levels of health, and well-being in many regions of the world (Navarro 1999).

Climate change is another crucial determining factor for health. Scholars and institutions have started to recognise the interconnection between climate change and global health, both in the sense that climate change is caused by humans but also the other way around, that climate change poses a major risk to human health (Sunyer and Grimalt 2006). 'It is now widely accepted that human activity (population growth and consumption patterns) contributes to climate change, environmental degradation and loss of biological diversity, all of which are already posing profound risks to health and all life that disproportionately affect the poorest, as exemplified by the emergence and spread of new zoonotic disease' (Benatar 2016, 602).

Global South citizens are double burdened as they are more vulnerable to the effects of climate change, while they also have the poorest health status. Further, the expenses to combat these challenges are unmanageable for low-income countries (Marino and Ribot 2012). The 'growth mantra' of the neoliberal paradigm has spread to the Global South that has been adopting a capitalist pattern of growth and development, which increases the global ecological footprint and the global burden of 'privileged' health issues like chronic diseases (Baum 2009). Therefore, the contemporary alteration of capitalist thinking towards a more holistic approach is much needed, as it gives Global South nations an opportunity to integrate sustainable initiatives at an early stage in their socioeconomic development.

In times of crisis, like the recent Covid-19 outbreak, neocolonialist tendencies are exacerbated. The pandemic highlights the inequity in health, access to health care, and the structural oppression of the marginalised, disadvantaged majority caused by Eurocentric power structures (Büyüm et al. 2020). The intricately colonial nature of the global health community is illustrated in Western governance demanding that nations in the Global South apply the same measurements in preventing the virus from spreading. In a still asymmetrical world, a majority of the Western population have the resources to not only survive restrictions but also manage quite well, while many in the Global South are facing problems of survival; this highlights the existing global inequalities (*ibid.*). Relevant to note is how some Global South countries, such as India, are able to produce and export vaccines, pointing to growing agency (Thiagarajan 2021). With the challenges to health equity as seen in social determinants, climate change, and Covid-19, '[T]there is emerging consensus about the need to decolonize Global Health' (Affun-Adegbulu and Adegbulu 2020). This decolonisation should occur on both epistemic and ontological levels, altering both the content, and the terms on which we are having the conversation (*ibid.*). Decolonisation must create equity within and across national borders and nationalities while recognising and reflecting the diversity of countries and people that are shaped by the power structures in the interface 'of historical, economical, social and political forces' (Büyüm et al. 2020, 1). Decolonising on a praxis-level is laid out in the section below, which zooms in on the individual, human encounter between North and South.

The human encounter between the Global North and the Global South

This section focuses on the human aspects of the encounters, usually overlooked in the mainstream global health literature, invoking both the agents of intervention and the targets. The postcolonial entry point seeks to create an understanding of the current condition, the 'human existential situation which we have often no power of choosing' (Tlostanova 2019, 165). We present studies within postcolonial frameworks in order to point to the historical and human consequences of colonialism, which serves as the backbone to launch the agency to change it. We engage and advocate decolonial praxis approaches as the 'conscious choice of how to interpret reality and how to act upon it' (*ibid.*).

The starting point is problematising the way agents of global interventions from the Global North are addressed as 'expatriates', moving to their own motives, frictions and moral dilemmas within the notion of 'doing good' while 'living well', through relevant empirical studies (Fechter 2012; Grover 2018a; Schliewe 2019). Finally, frameworks that include constructive strategies for fruitful encounters with emphasis on ethical justification, pragmatic, participatory approaches, and nuanced awareness about colonial legacies and global inequity (Easterly 2007; Singla 2012; Singla and Rasmussen 2018) are presented. These form the foundation for the social psychological dynamics, especially relational trust/mistrust (Greene et al. 2013; Keller 2006) along with ruptures.

The term expatriate entails colonial thinking which raises questions about its relevance in the contemporary world. This 'superior' social identity reinforces power relations and international imbalances in the Global South (Fechter 2011). In this context, the question is raised by Mawuna Remarque Koutonin (2015), a South African activist: 'Why are white people expats when the rest of us are immigrants?'. He answers, that every individual working abroad is an expat, regardless of skin colour and origin as the word comes from the Latin terms ex ('out of') and patria (country). In reality however, the term expat is reserved exclusively for Western white people. This also applies to Global North health professionals working in the Global South, which enables a hierarchical system where they enjoy superior positions with privileges attached. Koutonin (2015) thus appeals: 'If you see those "expats" in Africa, call them immigrants like everyone else. If that hurts their white superiority, they can jump in the air and stay there. The political deconstruction of this outdated worldview must continue' (*ibid.*, 2). Similarly, 'who is an expatriate in India?' is raised by Indian–Norwegian anthropologist Shalini Grover (Grover 2018a; 2018b). She concludes that expat is a term used for Euro-Americans among many Indians and points towards their privileged race and class reflecting postcolonial continuities, especially a continuation of persistent inequality. She also challenges the term, which resonates with the critique of the 'categorical fetishism' in migration research (Crawley and Skleparis 2018), that seeks to classify only certain persons as 'migrants' while others are treated as part of the cosmopolitan elites or expatriates. In line with this critique, we highlight Crawley and Skleparis' (2018) appeal to engage with the politics of bounding, the process by which such categories are constructed, their purpose and consequences, in order to challenge their use as a mechanism to divide and discriminate.

Nevertheless it is relevant to note breaks in the positioning of the previous 'colonised and the colonisers' (Korpela 2012) such as children of Asian migrants from the Global North who work in the Global South as health professionals. An example is Chinese origin researcher Lu Gram, raised in Denmark, educated in the United Kingdom, promoting women's and children's health through community groups in low-income and middle-income countries, now based in Mumbai (Gram et al. 2019). Is Gram an expat? Not according to the hegemonic discourse in which the term expat is reserved for Global Northern/white people, leaving them in a superior position—and both the Global North and South has a responsibility to change this hierarchical perception, in line with Koutonin's proposal. Through engaging in decolonising, we can challenge this narrow concept 'expat' and consider any person going to work outside their country an immigrant.

Now we move beyond the particular heritage and specific colonial cultures and direct our attention to the motives and dilemmas of the mobile health professionals as they are governed by various rules and policies, posed by the sending, Global North, and receiving Global South states (Korpela and Guha 2020).

Health professionals' mixed motives and moral dilemmas

German-British researcher Fechter (2012) examines health professionals' motivations through an empirical study of privileged migrants' mixed motivations based on a long-term exploration in Cambodia. She concludes that the desire to help and make a difference exists along with motivations for financial, professional, and personal benefits. However, it does not diminish their altruistic intentions and efforts. Altruism commonly means 'to live for others' (*vivre pour atrium*) and refers to actions that benefit others and not the agent. Historically, '...altruism represented the powerful impulse to the intellectual and moral development of humanity to which we must strive as a future state' (Mangone 2020). While considering the health professionals encounter with the other, it is important to take into account the diversity of mixed motives, which do not position them as either heroes, implying 'purely altruistic' motives, especially in Western popular imagination, nor villains, being excessively critical in line with Hancock's classical book 'The Lords of Poverty: The Power, Prestige, and Corruption of the International Aid Business' (Hancock 1992).

At the same time, we have to be aware of the moral dilemmas for perceiving the psychosocial situation of the health professionals. Such dilemmas faced by Global North professionals in the Global South are illustrated in the Danish scholar Sanna Schliewe's empirical studies (Schliewe 2017; 2019). A concrete example is that despite the 'Danish ideology' of equality, these professionals experienced 'loving that cheap-ironed shirt' by the local domestic staff who are part of an unregulated labour market. The hiring of cheap domestic staff implying inequality and inhumane working conditions enables expatriate families to lead comfortable lives, liberating them from onerous domestic chores. A nuanced longitudinal process of changes indicates that Global North professionals transform over time as they accept many novel practices related to domestic staff that they first found so difficult, especially the uncomfortable images of an exploitative master and servant relationship. This exemplifies the challenge to true decolonisation that is moving beyond decolonising thoughts to also decolonising practices, addressing the gap between thinking/saying and doing.

Similarly, a study by Shalini Grover (2018b) points out that the salaries Euro-Americans pay their domestic staff are akin to 'cheap labour' and incongruous with domestic workers' advanced skills of all-round domesticity. Still, treating domestic staff and endorsing affluent lifestyles seem like 'an exception', and thus ease its legitimisation and justify the inequity involved through sharing with other foreigners in 'expat Bubbles', also implying rather limited contact with the local population (Grover and Schliewe 2020; Schliewe 2019). These aspects of inequity in terms of wages and other forms of discrimination towards the local population are part of everyday life for Global North professionals in the South. We underline that these dilemmas, exploitations, and frictions have to be reflected upon and taken into consideration when planning health interventions from the Global North in Global South settings.

Inclusion of the perspective of the local other(s)

Besides historical aspects, current perspectives on the targets of the intervention, that is, the local population should be taken into consideration especially within the decolonial framework, in which there is an accentuation on indigeneity. In the practice of global health, there are many examples of experts from the Global North, executing health interventions in local settings in the Global South. The former Danida (The Danish International Development Agency) advisor, Jens Dalsgaard (2020) points out that most development experts and advisors lack field experience and the reality of local Global South communities; thereby it becomes distant and intangible. He argues that only few had development work as such under the skin and missed the direct, experiential sensory experience from the field.

We perceive that colonial legacies and neoliberal tendencies have affected the entire system of global health, including health professionals on a praxis level. While health professionals cannot change the whole system, they can contribute by being conscious about their own role in the matter. It is important that the health professionals confront their own background and position to ensure inclusion of the local context, local knowledge and the local 'other'. However, while inclusion of the local is important, at a more specific level, the ethical aspects must be addressed. 'Ethical justification'—ensuring the safety and well-being of the persons involved in health promotion intervention is salient. The Tuskegee syphilis study 1932–1972 (originally titled 'Tuskegee Study of Untreated Syphilis in the Negro Male') in United States of America serves as an example of unethical practice for a generation of practitioners and researchers (Tuskegee University 2020). Although the study was conducted in the Global North, the ethical foundations were shaky as the African–American participants were never told about or offered informed consent during the 40 years of the study.

To prepare health professionals going from the Global North to the Global South, an example can be found in a Danish University context with the initiation of a pragmatic interdisciplinary course, attempting to create a balance between the expert knowledge position and local knowledge systems through a multi-contextual combination of theory and practice (Singla and Rasmussen 2018). The course includes an emphasis on the significance of first-hand experience and inclusion of the motives and dilemmas of the agents of intervention, which have been overlooked in the classical training literature (Ward, Bochner, and Furnham 2002). According to Easterly (2007) pragmatic aspects such as the basic principle of accountability and evaluation based on symmetrical involvement and feedback from the intended beneficiaries should be emphasised.

Decolonisation of health promotion praxis

In working with the aforementioned supersetting approach, it is sensible to introduce decolonising praxis frameworks. The Applied Decolonial Framework for Health Promotion is developed by Paul Chandanabhumma and Subasri Narasimhan (2020) who define decolonisation as an active, calculated resistance to the forces of colonialism, that perpetuate the subjugation of mind, bodies, and lands with the purpose of overturning the colonial structure and realising indigenous liberation and aims at integrating decolonial processes into health promotion practices. They emphasise how '[T] the framework will help health promotion stakeholders attend to colonising structures within the field and engage with communities to achieve social justice and health equity' (*ibid.*, 831).

The framework consists of three overlapping domains; reflection, planning, and action, which combined leads to engagement in decolonial processes. The reflection domain concerns decolonising one's mind through three sub-elements: '…assessing long-term effects of colonisation, critique of current paradigms and decentering frame of reference' (*ibid.*, 833). The planning domain involves mutual dialogue based on trust, respect, and transparency while foregrounding local knowledge. In the action domain, the nature of dialogue is imperative for building alliances across groups, engaging in social justice, and placing self-determination at the core of activities to maintain commitment to minimise and/or eliminate the colonial legacy. The framework seeks to encourage health professionals to break with colonial legacies that continuously define power structures while aiding marginalised groups out of oppression.

A similar rationality can be found in Two-Eyed Seeing Relational Process presented through a research project on diabetes prevention within an Indigenous community in Quebec Canada by Hovey et al. (2017). The project demonstrated a Global North and Global South health promotion research partnership. As the name suggests, the process involves 'seeing together', and thus evokes equity for the stakeholders involved. Based on a principle of shared ownership of knowledge, the process involves a mutual respect and recognition that there exist multiple worldviews, and a plurality of understandings of the same things, among them healthy lives and communities (*ibid.*). In opposition to the ways in which Western knowledge tends to need placement within a specific discipline or area, the Two-Eyed Seeing Relational Process is concerned with life in ways that cannot be categorised. In the same manner as the Applied Decolonial Framework for Health Promotion, the Two-Eyed Seeing Relational Process may allow us to decolonise thoughts and approaches to global health promotion practices in a local context. Its purpose is to create a foundation for shared ownership of knowledge that includes perspectives on both sides of the encounter; new knowledge that can ultimately become part of the global health promotion practices.

It is vital to note that Chandanabhumma and Narasimhan (2020) discuss a pragmatic and problematic issue related to inclusion of indigenous wisdom and co-option of this wisdom by the larger society (e.g. for commercial value) with little benefit to the local population. Concretely, they appeal for forms of knowledge through acknowledging indigenous as a form of intellectual property requiring attribution, co-participation, and remuneration for knowledge sharing, exercising elements of action, so that decolonising frameworks are not just an intellectual exercise but actually advances systemic changes. This resonates well with British—Sri Lankan psychiatrist Suman Fernando's (2014) appeal regarding global mental health care, acknowledging that there is much wisdom in nonwestern countries and we can learn from how other cultures handle humanitarian problems.

In most global contexts pragmatic power dynamics and commercial values conflicts have to be addressed and actively attended to, as it is not possible to work fruitfully within a traditional setting, without being aware of, open to, and working in consonance with the local social dynamics (Chaudhary 2008; Dutta 2008; Fernando 2014). At the pragmatic level, we also direct attention to the major barriers related to decolonisation and local inclusion such as collective amnesia caused by traumatic memories, difficulties in translating community agency to active resistance, and actual power differential dynamics and sharing. Thus, community–academic–activist connections that sustain community-led interventions for health and well-being are seen as ways of addressing these barriers and moving forward.

Health professionals engaging in a decolonising praxis should be mindful of using their professional skills in creating and supporting a healthier environment, while ensuring that locals have a voice in the matter of their own lives and well-being; thus, ensuring a sense of ownership for the individuals involved. It is crucial to include the local's voice and praxis of their own epistemologies not only to ensure relevance for themselves, but more importantly because the Global South for so long has not had their own voice present in the development of global health. Thus, it is significant that decolonising the field requires the creation of a new space where practitioners privilege the theories, values, and interests of marginalised communities, while at the same time being mindful of the imposition of external interests. Indian psychologist Ashis Nandy (1983) delineates loss and recovery of the self under colonialism, including decolonisation of the mind. He emphasises that coloniality patterns are not only about economic or technological domination, but cultural subservience of indigenous peoples and the cultural arrogance of the power elite; while the Gandhian movement in India could be partly understood as an attempt to transcend a strong tendency of educated Indians to articulate political striving for independence in European terms. However, coloniality here deals with and confronts issues and problems common to all former colonies of Western Europe in the 'third World' in its history of more than 500 years. The coloniality of power (Mignolo and Walsh 2018) operates today on a global scale when North Atlantic imperial states can no longer control and are disputed by emerging economies. These processes can be seen as some of the creating and illuminating pluriversal and interversal paths that disturb the totality from which the universal and the global are perceived (*ibid.*).

While interventions executed in an empowering and participatory manner are comprehensive and time demanding, by combining the supersetting approach with the two presented decolonising frameworks, health promoting professionals may contribute to the sustainable development of global health promotion practices (Bloch et al. 2014). Surely the combined framework does not only apply to the Global South but to the global community in modern times, regardless of the local context. Through an innovative approach to health promotion based on globally shared visions, we can assign health professionals the power to start changing the '...behavioural, ideological, institutional, political, economic and cultural systems...' (Chandanabhumma and Narasimhan 2020, 831).

We end this section with an appeal to foster enduring, strong, genuine, horizontal collaborative, and trusting international alliances in the field of global health promotion (Chaudhary and Sriram 2020). Such cooperation should also take into account the current context, status, motives, and dilemmas of health professionals, along with the involvement of local populations through a decolonising mind-set with the ultimate purpose of altering the concept of global health so that it keeps its implicit promise of being by and for the entire global population.

Concluding remarks: global health by and for the global community

We have pointed to how inequity in global health is created through the historical development of the field and sustained through the societal structure. The current neoliberalist order is built on a capitalist pattern that structurally leads to inequities in socioeconomic status. Health is deprioritised as the development agenda is concentrated on economic growth and capital is therefore economically defined. As highlighted by other scholars, we convey that the terms of the current global health paradigm are failing to adequately address the contemporary effects of colonial legacies and the unequal power structures resulting in the systemic oppression of marginalised nations (Chandanabhumma and Narasimhan 2020). In current societal structures, global threats, for example, the Covid-19 pandemic, create a risk for protectionist tendencies, which might make nations increasingly turn inwards worrying about national health and security—a danger to global cooperation and the quest for global equity and health in the future (Schwab 2020). For this reason and because the current neoliberalist and capitalist structures counteract sustainable, healthy development, it makes sense that an increasing number of scholars are pointing to the need for a paradigm shift entailing a more holistic approach to sustainable global development. Büyüm et al. (2020, 3) suggest a paradigm shift that repoliticises global health 'by grounding it in a health justice framework that acknowledges how colonialism, racism, sexism, capitalism and other harmful '-isms' pose the largest threat to health equity'. Schwab (2020) suggests a need to include health, equity, and sustainability in the definition of capital. As such, the concept of global health needs to be extended to holistically include the context of surrounding factors that economically, socially, and environmentally may influence global health and vice versa. As highlighted by Baum (2009), a more equal distribution of ecological footprints sets the stage for health equity.

By disassembling constituting elements of global health, it becomes clear how the contemporary field is based on and biased in Global North cultures and traditions, which frames the Eurocentric perspective as the neutral and the norm, and therefore solutions and strategies for combating global risks are customised to fit societies of the Global North. With identified pathways towards equity in global health, we suggest focusing the decolonisation process on the four different levels presented in the introduction: *knowledge, leadership, policy, and praxis.* To decolonise *knowledge* production is to confront the fact that scientific knowledge is predominantly produced by the Global North. By doing so, global health science can be situated and understood for what it is, merely a perspective. As pointed out in recent literature on the matter '…who is generating the knowledge, and why they are in a position to do so—matter profoundly'(Brisbois, Spiegel, and Harris 2019, 3). To accommodate the health challenges of the world as one global community, all members of the community ought to be included on equal terms. This inclusion also applies to representation at the decision-making table. Members of the Global South ought to be part of the global health *leadership* and hereby have equal influence on the global health agenda (Büyüm et al. 2020); as such we encourage Global North leaders to take responsibility in making room for the Global South. If global health institutions continue to lack representation of Global South nations at the decision-making table, policies, and frameworks will continue to mirror a Eurocentric worldview, and thus not adequately consider the majority of the world's needs (*ibid.*). Hereby, equal collaboration in global health partnerships also entails initiatives established and led by Global South countries, securing the inclusion of a Global South perspective in the global agenda. Contemporary *policies* and frameworks, including the SDGs, are perhaps the most important tools in working towards a better and more just world. However, it is important to understand that the foundation of the SDGs presume 'that the same economic system, and it's still-present neoliberal governing rules, that have created or accelerated our present era of rampaging inequality and environmental peril can somehow be harnessed to engineer the reverse' (McCoy 2017, 541). Regardless of the political and societal structures, good *praxis* considerations for global health promotion must include a holistic approach, involving tools of participation and co-creation to ensure relevance for the local population. Intervention frameworks intentionally should not put forward a completed solution model. In order to decolonise health promotion, we argue that pre-defined solutions should be replaced by approaches that are adapted specifically to local contexts and co-created with the involved communities and individuals. In working with global health we urge professionals to pay attention to not only decolonise thoughts but also practices. In doing so, we have challenged hegemonic categories such as 'expatriate' and the encounter between the health professional and the locals. Attention should be directed towards the difference between saying and practicing, where decolonial studies often highlight this gap as an important challenge to equity.

We want to encourage future scholars to actively engage in decolonising the field of global health, by further exploring the presented challenges. In working with the complex and ever evolving field of global health promotion we want to stress the importance of at all times keeping focus on the ethical consideration; that we all as human beings must be regarded as equals. It is therefore necessary for scholars and professionals to move away from universalism and towards decolonial pluriversalism in the field of global health; that is to acknowledge the multiple ways of being and doing that have always existed. Finally, we want to convey our main message; that global health promotion should be *by* and *for* all.

References

Affun-Adegbulu, Clara, and Opemiposi Adegbulu. 2020. "Decolonising Global (Public) Health: From Western Universalism to Global Pluriversalities." *BMJ Global Health* 5, no. 8: e002947. https://doi.org/10.1136/bmjgh-2020-002947.

Baum, Fran. 2009. "Envisioning a Healthy and Sustainable Future: Essential to Closing the Gap in a Generation." *Global Health Promotion IUHPE* 16, no. 1: 72–80.

Benatar, Solomon. 2016. "Politics, Power, Poverty and Global Health: Systems and Frames." *International Journal of Health Policy and Management* 5, no. 10: 599–604. https://doi.org/10.15171/ijhpm.2016.101.

Bloch, Paul, Ulla Toft, Helene Christine Reinbach, Laura Tolnov Clausen, Bent Egberg Mikkelsen, Kjeld Poulsen, and Bjarne Bruun Jensen. 2014. "Revitalizing the Setting

Approach – Supersettings for Sustainable Impact in Community Health Promotion." *International Journal of Behavioral Nutrition and Physical Activity* 11, no. 1: 118. https://doi. org/10.1186/s12966-014-0118-8.

Bochkov, Danil. 2021. "Great Power Competition and the COVID-19 Vaccine Race." 2021.https://thediplomat.com/2021/01/great-power-competition-and-the-covid-19-vaccine-race/.

Brisbois, Ben W., Jerry M. Spiegel, and Leila Harris. 2019. "Health, Environment and Colonial Legacies: Situating the Science of Pesticides, Bananas and Bodies in Ecuador." *Social Science & Medicine* 239 (October): 112529. https://doi.org/10.1016/j. socscimed.2019.112529.

Büyüm, Ali Murad, Cordelia Kenney, Andrea Koris, Laura Mkumba, and Yadurshini Raveendran. 2020. "Decolonising Global Health: If Not Now, When?" *BMJ Global Health* 5, no. 8: e003394. https://doi.org/10.1136/bmjgh-2020-003394.

Chakrabarti, Pratik. 2014. "Conclusion: The Colonial Legacies of Global Health." 200–205. https://doi.org/10.1007/978-1-137-37480-6_11.

Chandanabhumma, P. Paul, and Subasri Narasimhan. 2020. "Towards Health Equity and Social Justice: An Applied Framework of Decolonization in Health Promotion." *Health Promotion International* 35, no. 4: 831–840. https://doi.org/10.1093/heapro/ daz053.

Chaudhary, Nandita, 2008. "Methods for a Cultural Science." *Constructing Research Methods: Insights from the Field* 50: 29–52. New Delhi: SAGE.

Chaudhary, Nandita, and Sujata Sriram. 2020. "Psychology in the 'Backyards of the World': Experiences from India." *Journal of Cross-Cultural Psychology* 51, no. 2: 113–133. https://doi.org/10.1177/0022022119896652.

Crawley, Heaven, and Dimitris Skleparis. 2018. "Refugees, Migrants, Neither, Both: Categorical Fetishism and the Politics of Bounding in Europe's 'Migration Crisis.'" *Journal of Ethnic and Migration Studies* 44, no. 1: 48–64. https://doi.org/10.1080/13691 83X.2017.1348224.

Dalsgaard, Jens Peter Tang. 2020. "Tidligere Danida Rådgiver: Jeg Stoppede Efter Ti År— Med Ondt i Hovedet Og Samvittigheden." Altinget.Dk (blog). 2020. //www.altinget. dk/artikel/tidligere-danidaraadgiver-jeg-stoppede-efter-ti-aar-med-ondt-i-hovedet-og-samvittigheden. Accessed August 27, 2021.

Dixey, Rachael, Louise Warwick-Booth, Jane South, Diane Lowcock, Ivy O'Neil, Judy White, and James Woodall. 2013. "Healthy Public Policy." Ed. Rachael Dixey In *Health Promotion: Global Principles and Practice*, 54–77. Modular Texts. Wallingford, Oxfordshire: CABI.

Dutta, Mohan J. 2008. "Health, Culture and Globalization." In *Communicating Health: A Culture Centered Approach*, 235–251. Ed. Mohan J. Dutta Cambridge: Polity Press.

Easterly, William. 2007. *The White Man's Burden: Why the West's Efforts to Aid the Rest Have Done So Much Ill and So Little Good. OUP Catalogue.* New Delhi: Oxford University Press.

Fechter, Anne-Meike. 2011. "Anybody at Home? The Inhabitants of 'Aidland.'" Eds. Anne-Meike Fechter & Heather Hindman In *Inside the Everyday Lives of Development Workers: The Challenges & Futures of Aidland.* Sterling, VA: Kumarian Press.

Fechter, Anne-Meike. 2012. "'Living Well' While 'Doing Good'? (Missing) Debates on Altruism and Professionalism in Aid Work." *Third World Quarterly* 33, no. 8: 1475– 1491. https://doi.org/10.1080/09700161.2012.698133.

Fernando, Suman. 2014. *Mental Health Worldwide: Culture, Globalization and Development.* Basingstoke: Palgrave Macmillan.

Frederiksen, Mette. 2020. "Mette Frederiksens tale ved pressemøde om nedlukning mellem jul og nytår 2020." Press conference speech presented at the Press conference on Covid-19, Statsministeriet, Denmark, December 16.

Gram, Lu, Adam Fitchett, Asma Ashraf, Nayreen Daruwalla, and David Osrin. 2019. "Promoting Women's and Children's Health through Community Groups in Low-Income and Middle-Income Countries: A Mixed-Methods Systematic Review of Mechanisms, Enablers and Barriers." *BMJ Global Health* 4, no. 6: e001972. https://doi.org/10.1136/bmjgh-2019-001972.

Greene, Jeremy, Marguerite Thorp Basilico, Heidi Kim, and Paul Farmer. 2013. "Colonial Medicine and Its Legacies." Eds. Paul Farmer, Jim Yong Kim, Arthur Kleinman & Matthew Basilico In *Reimagining Global Health*, 33–76. Berkerley: University of California Press.

Grosfoguel, Ramón. 2011. "Decolonizing Post-Colonial Studies and Paradigms of Political-Economy: Transmodernity, Decolonial Thinking, and Global Coloniality." *TRANSMODERNITY: Journal of Peripheral Cultural Production of the Luso-Hispanic World* 1, no. 1: 1–38. Permalink: https://escholarship.org/uc/item/21k6t3fq

Grover, Shalini. 2018a. "Who Is an Expatriate? Euro-American Identities, Race and Integration in Postcolonial India: New Managerial Roles, Social Mobility and Persistent Inequality." Eds. Sanna Schliewe, Nandita Chaudhary & Giuseppina Marsico In *Cultural Psychology of Intervention in the Globalized World*, 283–295. Advances in Cultural Psychology: Constructing Human Development. Charlotte, NC: Information Age Publishing.

Grover, Shalini. 2018b. "English-Speaking and Educated Female Domestic Workers in Contemporary India: New Managerial Roles, Social Mobility and Persistent Inequality." *Journal of South Asian Development* 13, no. 2: 186–209. https://doi.org/10.1177/0973174118788008.

Grover, Shalini, and Sanna Schliewe. 2020. "Trailing Spouses (India)." In *Global Encyclopedia of Informality*. https://www.in-formality.com/wiki/index.php?title=Trailing_spouses_(India).

Hancock, Graham. 1992. *The Lords of Poverty: The Power, Prestige, and Corruption of the International Aid Business*. New York: Atlantic Monthly Press.

Horton, Richard. 2019. "Offline: Transcending the Guilt of Global Health." *The Lancet* 394, no. 10203: 996. https://doi.org/10.1016/S0140-6736(19)32177-4.

Hovey, Richard, Treena Delormier, Alex McComber, Lucie Lévesque, and Debbie Martin. 2017. "Enhancing Indigenous Health Promotion Research through Two-Eyed Seeing: A Hermeneutic Relational Process." *Qualitative Health Research* 27 (March): 104973231769794. https://doi.org/10.1177/1049732317697948.

Hussain, Sameera, Dena Javadi, Jean Andrey, Abdul Ghaffar, and Ronald Labonté. 2020. "Health Intersectoralism in the Sustainable Development Goal Era: From Theory to Practice." *Globalization and Health* 16, no 1: 15. https://doi.org/10.1186/s12992-020-0543-1.

Keller, R.C. 2006. "Geographies or Power, Legacies of Mistrust: Colonial Medicine in the Global Present." *Historical Geography* 34 (January): 26–48.

Korpela, M. (ed.). 2012. "A Postcolonial Imagination? Westerners Searching for Authencity in India." In *The New Expatriates: Postcolonial Approaches to Mobile Professionals*, 1st edition, 91–109. London: Routledge.

Korpela, M., and P. Guha. 2020. "CfP—Expat Families: Rules, Power, Participation and Transgression." In Helsinki. https://www.siefhome.org/congresses/sief2021/cfpan.

Kothari, Ashish, Ariel Salleh, Arturo Escobar, Federico Demaria, and Alberto Acosta (eds.). 2019. *Pluriverse: A Post-Development Dictionary*. New Delhi: Tulika Books.

Koutonin, Mawuna Remarque. 2015. "Why Are White People Expats When the Rest of Us Are Immigrants?" *The Guardian*. Working in Development. http://www.theguardian.com/global-development-professionals-network/2015/mar/13/white-people-expats-immigrants-migration.

Mangone, Emiliana. 2020. *Beyond the Dichotomy Between Altruism and Egoism: Society, Relationship, and Responsibility*. History and Society: Integrating Social, Political and Economic Sciences. Charlotte, NC: Information Age Pub Inc.

Marino, Elizabeth, and Jesse Ribot. 2012. "Special Issue Introduction: Adding Insult to Injury: Climate Change and the Inequities of Climate Intervention." *Global Environmental Change* 22, no. 2: 323–328. https://doi.org/10.1016/j.gloenvcha.2012.03.001.

Marmot, Michael. 2005. "Social Determinants of Health Inequalities." *Public Health* 365: 6.

McCoy, David. 2017. "Critical Global Health: Responding to Poverty, Inequality and Climate Change Comment on 'Politics, Power, Poverty and Global Health: Systems and Frames.'" *International Journal of Health Policy and Management* 6, no. 9: 539–541. https://doi.org/10.15171/ijhpm.2016.157.

McCracken, Kevin, and David Rosser Phillips. 2012. "Concepts, Data, Measurements, and Explanations." Eds. McCracken, Kevin, and David Rosser Phillips In *Global Health: An Introduction to Current and Future Trends*, 3–22. London, UK: Routledge.

Mignolo, Walter, and Catherine E. Walsh. 2018. *On Decoloniality: Concepts, Analytics, Praxis*. On Decoloniality. Durham: Duke University Press.

Navarro, Vicente. 1999. "Health and Equity in the World in the Era of 'Globalization.'" *International Journal of Health Services* 29, no. 2: 215–226. https://doi.org/10.2190/MQPT-RLTH-KUPJ-2FQP.

Nandy, Ashis. 1983. *The Intimate Enemy: Loss and Recovery of Self Under Colonialism*. Delhi: Oxford.

Reidpath, Daniel D., and Pascale Allotey. 2019. "The Problem of 'Trickle-down Science' from the Global North to the Global South." *BMJ Global Health* 4, no. 4: e001719. https://doi.org/10.1136/bmjgh-2019-001719.

Richey, Lisa Ann. 2015. *Celebrity Humanitarianism and North-South Relations: Politics, Place and Power*. Abingdon, Oxon: Routledge.

Saha, Sudipta, Purvaja Kavattur, and Amina Goheer. 2019. "The C-Word: Tackling the Enduring Legacy of Colonialism in Global Health." *Health Systems Global*. https://healthsystemsglobal.org/news/the-c-word-tackling-the-enduring-legacy-of-colonialism-in-global-health/.

Schliewe, Sanna. 2017. "Resisting Inequality but Loving Those Cheap Ironed Shirts: Danish Expatriates' Experiences of Becoming Employers of Domestic Staff in India." In *Resistance in Everyday Life: Constructing Cultural Experiences*, 181–201. Singapore: Springer. https://doi.org/10.1007/978-981-10-3581-4_14.

Schliewe, Sanna. 2019. "Uneasy Encounters and Privileged Migration: A Cultural Psychology Study of Danish Expatriates and Their Domestic Workers in India." Ph.D., Center for Cultural Psychology, Aalborg: Aalborg University Denmark.

Schwab, Klaus. 2020. "Post-COVID Capitalism." Project Syndicate. October 12, 2020. https://www.project-syndicate.org/commentary/post-covid-capitalism-great-reset-by-klaus-schwab-2020-10.

Singla, Rashmi. 2012. "Migration, etnisk diversitet og sundhedsfremme." Eds. Betina Dybbroe, Birgit Land & Steen Baagøe Nielsen In *Sundhedsfremme: et kritisk perspektiv (Health promotion: A critical perspective)*, 1. udgave. Frederiksberg: Samfundslitteratur.

Singla, Rashmi, and Louise Mubanda Rasmussen. 2018. "Global Health Intervention from Global North to Global South: (Academic) Preparation of Students in Cultural Psychology of Intervention in the Globalized World." Eds. Sanna Schliewe, Nandita Chaudhary & Giuseppina Marsico In *Cultural Psychology of Intervention in the Globalized World*, 257–287. Advances in Cultural Psychology: Constructing Human Development. Charlotte, NC: Information Age Publishing.

Stewart, Kearsley A., and Kelley K. Swain. 2016. "Global Health Humanities: Defining an Emerging Field." *The Lancet* 388 (10060): 2586–2587. https://doi.org/10.1016/S0140-6736(16)32229-2.

Sunyer, Jordi, and Joan Grimalt. 2006. "Global Climate Change, Widening Health Inequalities, and Epidemiology." *International Journal of Epidemiology* 35, no. 2: 213–216. https://doi.org/10.1093/ije/dyl025.

Thiagarajan, Kamala. 2021. "Covid-19: India Is at Centre of Global Vaccine Manufacturing, but Opacity Threatens Public Trust." *BMJ* 372 (January): 196. https://doi.org/10.1136/bmj.n196.

Tlostanova, Madina. 2019. "The Postcolonial Condition, the Decolonial Option, and the Post-Socialist Intervention." In *Postcolonialism Cross-Examined: Multidirectional Perspectives on Imperial and Colonial Pasts and the Neocolonial Present*, 165–178. Routledge. https://doi.org/10.4324/9780367222543-9.

Tlostanova, Madina, Suruchi Thapar-Björkert, and Ina Knobblock. 2019. "Do We Need Decolonial Feminism in Sweden?" *NORA—Nordic Journal of Feminist and Gender Research* 27, no. 4: 290–295. https://doi.org/10.1080/08038740.2019.1641552.

Tuskegee University. 2020. "About the USPHS Syphilis Study." 2020. https://www.tuskegee.edu/about-us/centers-of-excellence/bioethics-center/about-the-usphs-syphilis-study.

Ward, Colleen A., Stephen Bochner, and Adrian Furnham. 2002. "Culture Training." In *The Psychology of Culture Shock*, 245–265. East Sussex: Routledge.

WHO (ed.). 2005. *Health and the Millennium Development Goals*. Geneva, Switzerland: World Health Organization.

WHO. 2009. "Milestones in Health Promotion: Statements from Global Conferences." World Health Organization.

WHO. 2016. "Health Promotion." WHO. World Health Organization. http://www.who.int/healthpromotion/fact-sheet/en/.

WHO. 2020. "WHO | Decentralisation." WHO. World Health Organization. http://www.who.int/health-laws/topics/governance-decentralisation/en/.

WHO. n.d. "Social Determinants of Health." Accessed December 12, 2020. https://www.who.int/westernpacific/health-topics/social-determinants-of-health.

4 Theorising water, shifting scales

The space of the Himalayan Anthropocene

Prem Poddar

Introduction: future of water

...water, by assuming different forms, becomes this earth, sky, heaven, mountains, gods and men, cattle, birds, herbs and trees, all beasts down to worms, midges, and ants. Water itself assumes all these forms. Meditate on water (Chāndogya Upaniṣad 7.10.1).

(Translated in Mueller 1897, 177)

Great One is the honorific style (zi 字) given to water. It was first the mother of the sky and earth, and later the fount (yuan 源) of the myriad living things.

(Quoted in Yao Zhihua 2014, 5)[1]

this is the Koshī and that is the Gandak,
the blue over there is the Kanrālī.
You may have read in some papers
about the selling of Nepal's rivers.
That was a lie, sir.
Those rivers have given our regions their names,
we plan to generate power from them:
could you give us some help?

(Bishta 1991, 157)

This introductory section provides a brief analysis of two science fiction novels from South Asia to highlight what may be the state of play in the future in terms of politics of water as well as the representations of hydrological crisis. The segment allows me to consider the present significance of water in a way that we do not ponder over it or deliberate it in our everyday lives: to link it to historical concerns and also to ask what it may mean to think of it as a special object in order to consider later in this chapter what the era of the Anthropocene means for us and how we critically, artistically, and affectively approach objects such as water. The two epigraphs, one from ancient India and the other from ancient China, also provide a more philosophical

DOI: 10.4324/9781003172413-5

meditation on the nature of water while gesturing to the intimate interconnectivity of its flow within everything that is living as well as the non-human.
While the *Chāndogya Upaniṣad* from, circa, the 8th–6th century B.C.E.,
emphasises that water assumes different myriad forms, in Laozi (also rendered
as Lao Tzu, ancient Chinese philosopher and writer) from circa, 4th–6th
B.C.E. water is similarly the font and fount, and no less than the mother of
the sky and earth. The third epigraph is from the Himalayan state of Nepal,
and is a satirical reflection on the scaredness of the national in relation to the
economy of hydroelectric power generation and how the rivers of Nepal, and
the country itself, is subject to international aid. These texts bring forward
issues, especially from three Himalayan states, in relation to water that are
simultaneously in the past, present, and future and pose a pressing question
that is bigger than just their national economies which are ultimately dependent on a plentiful supply of water as are the lives and well-being of their
societies.

In Ian McDonald's acclaimed science fiction novel, *River of Gods* (2004),
portraying India in 2047, water wars rage across the country. One of the
more desperate states (in a now disunited India) has plans of towing icebergs
from what's left of the Antarctic ice sheet into the Bay of Bengal to the mouth
of the Ganga, to overcome the freshwater shortage but mostly in the hopes of
kick-starting long-delayed monsoons that are subject to the proper functioning of the hydro cycle linking the Indian ocean with the Himalayas.[2]

Since the monsoon failed, water; by the bottle, by the cup, by the sip, from
tankers and tanks and shrink-wrapped pallets and plastic litrejohns and back-
packs and goatskin sacks. Those Banglas with their iceberg, you think they'll
give us one drop here in Bharat? Buy and drink (Mcdonald 2004, 3).

Imagined thus, in this future India which has become Balkanized into
a number of smaller competing states, such as Awadh, Bharat, and Bangla,
water has become a social and political issue. The global information network, in many ways controlling it, is now inhabited by artificial intelligences,
phonetically called *aeais* in the novel. The novel is structured in such a way
as to follow a number of different characters' viewpoints on and around the
date of August 15, 2047. The date is remarkable as it denotes the centenary
of India's partition and independence from the British Raj. It is not as if the
novel is a direct critique of colonialism and the effects of its violence on this
future India. It is, in my reading, a world that Dipesh Chakraborty advocates,
where a simplistic postcolonial take—in which the installation of the carbon
economy worldwide is mainly the product of industrialising empires from
the Global North beginning at least in the 18th century, if not the colonial
conquest from the 16th century—sidestepping the climate emergency that
we face globally, planetarily, is not available.[3] In this sense the chapter is not
addressing head on the dominant ethic of colonialism that Anil Agarwal
and Sunita Narain in their influential paper 'Global warming in an unequal
world: a case for environmental colonialism' (1990) levied in charging that
the report of the World Resources Institute (WRI) in collaboration with UN

was 'based less on science and more on politically motivated and mathematical jugglery. Its main intention seems to be to blame developing countries for global warming and perpetuate the current global inequality in the use of the earth's environment and its resources' (Agarwal and Narain 1991, 1). The world has changed in these sci-fi novels but we are still in nation-state or region specific maps with power unevenly divided.

For now, it is imperative to delineate the water politics staged in the novel, and why this may be significant. The adviser for Bharat, Shaheen Badoor Khan, is charged by Bharat's Prime Minister with intra-nationally meeting and escorting his ministers to Bangla.

> Shaheen Badoor Khan looks down on to the Antarctic ice. From two thousand metres it is less ice than geography, a white island, Sri Lanka gone rogue. The ocean-going tugs hired from the Gulf are the biggest and strongest and newest but they look like spiders tackling a circus big top, hauling away at silk thread guy ropes. Their role is supervisory now; the Southwest Monsoon Current has the berg and the whole performance is running north-by-northeast at five nautical miles per day. Out here on the ocean five hundred kilometres south of the delta the only visual referents are ice and sky and the dark blue of deep water,...
> ...the States of Bengal tilt-jet lurches in the chill microclimate spiralling up from the ice floe. Shaheen Badoor Khan notices that the surface is grooved and furrowed with crevasses and ravines. Torrent water glitters; ice-melt has gouged sheer canyons in the ice walls, spectacular waterfalls arc from the berg's cliff edges. (22)

Shaheen Badoor Khan is clearly awe-struck but also feels 'deep shame' when one of his ministers exhibits his ignorance in front of the silver-tongued Banglas. But he has no doubt that politics is not about exceptional talent, intelligence, or skill—it is his job as advisor to provide that or at least the appearance of that. The Bengali 'Minister With Iceberg', cannot help being proud, in a gesture of extant postcolonial cringe, and crows that they 'didn't run to the Americans to sort out [their] water supply, like those ones in Awadh' (Mcdonald 2004, 23).

Shaheen is thus offered an opportunity to make a politically charged point, also a regional, ex-national postcolonial, and by extension a holistic one: 'The river used to make us one country,' Shaheen Badoor Khan observes. 'Now we seem to be the squabbling children of Mother Ganga; Awadh, Bharat, Bengal. Head, hands, and feet' (Mcdonald 2004, 23).

This decapitated body of limbs is suggestive of how even a sacred river like Ganga is partitioned, divided, and fought over, not different from the squabbles over global carbon trading and climate change responsibilities. 'It's constantly shifting', says the energetic Bangla climatologist across the aisle. 'As it loses mass, the centre of gravity moves. We have to maintain equilibrium, a sudden shift close in could prove catastrophic' (Mcdonald 2004, 22).

The language is vulnerable to being read as the ones routinely trotted out in climate change summits in terms of the political scape of action: talk of 'constantly shifting' and losing 'equilibrium' lends itself to a mimicry. Macdonald here is masterful in satirising the meeting and the official Bangla climatologist who is showing off his knowledge, 'It has all been worked out to the last gram,' he says. 'We are well within the parameters for microclimatic shift' (Mcdonald 2004, 23). It is not unlike climate change conferences where experts display their detailed knowhow of data and information, but may be missing the planetary urgency of the matter at hand.

The characters in the novel go on their daily lives, but what/who is also significant is Aj (or literally 'today', translated from Hindi), the waif, the mind reader, and the prophet, someone who represents the concerns of the subaltern. The characters, as the narrative unfolds, are swept together to decide the fate of the nations as the great river Ganges flows on. The novel, among its many themes, represents for our purposes here, a terrifyingly plausible trope of the near future: water wars between nations. The region of Bharat hurtling towards war with neighbouring Awadh is one scenario; what is more plausible above all is the frightening future history of water wars between India and China. Add Pakistan, Bangladesh, Nepal, and some South East Asian nations to get a heady mix.

In Ruchir Joshi's *The Last Jet-Engine Laugh* (2001), set in Calcutta in 2030, the novel invokes a set of problems, including environmental degradation and a water crisis. It imagines a dystopia of total privatisation of water, summarised in a refrain adopting the Ancient Mariner's lament, 'waterwater everywhere notadrop to drink'. Joshi's novel pictures how the expanding urban communities fight over freshwater. Water tankers on the way to gated communities are hijacked, and the urban 'water wars' intensify until they have reached the inter-state level by 2030. Pakistan and India get involved in an ecological warfare where spraying of poisonous chemicals over glacial meltwater is part of the exchanges.

The hazardous pollution of the water supply in the novel necessitates that all water-related activities can only be conducted through technological intervention. Specially designed and packaged water substitutes for hydration are in the market as are devices designed to protect body orifices from dangerous water-substituting liquids during bathing. The chief narrator, Paresh Bhatt is tormented by his inability to describe the sensory experience of consuming real water:

> I explain, I cannot explain (…) water, like this: first of all moving thing; next, if real water then no colour, not really, after air and even in the air, main moving thing that you can see that has no colour. Third, effect on tongue and top of mouth and inside of cheeks as if something coming home. Transformed into an eerily synthetic commodity, water has been stripped of its symbolic and somatic singularity.
>
> (Joshi 2001, 23)

Paresh is adrift and somewhat lost. His daughter Para is aggressive and martial, and is the antithesis of her grandparents' principles of Gandhian non-violence, civil disobedience and passive resistance. When India is at war with a Muslim Pakistani–Iranian alliance, Para is a squadron leader in the Indian Air Force.

This postcolonial trajectory of Indian history from non-violence to belligerent jingoism is also a portent of apocalyptic futures. *The Last Jet-Engine Laugh* presages of a potential future wherein privatisation of water commons and toxification on a massive scale has followed in the wake of an ecological unravelling of South Asia.

Climate concerns: water

Writing and thinking about the future is actually writing about our own present in that the question becomes: who has access to resources, who benefits from enclosing them? How do these material conditions sculpt daily consciousness and societal culture? Both the novels even though they are located in a somewhat Orwellian future, they are recognisably painting a picture of plausibility.

The implications of survival in a fictional ambience of apocalyptic despair, also underlines the urgency to focus upon the possibilities of hope for survival as seen in many films and other fictional representations in scenarios and circumstances of apparently unmitigated hopelessness. We can read the above two texts variously. Do the texts stress in some ways the 'dark ecology' (Morton 2010, 16), where the argument is that ecological awareness in the present Anthropocene era bears the structure of a 'strange Möbius strip'[4], twisted to have only one side as ecological phenomena are also fundamental to the make-up of how things are; leading to the idea that an idealized view of nature for Anthropecinists is not possible.[5] 'Deckard travels this oedipal path in *Blade Runner* (1982) when he learns that he might be the enemy he has been ordered to pursue. Ecological awareness takes this shape because ecological phenomena have a loop form that is also fundamental to the structure of how things are.' (Morton 2016, blurb). Or, is it that humans are subjected to displacement or condemned to habitation of a place stripped of the very values and properties that made it habitable at some time? (Nixon 2011, 19).

Whether it's China or India or any developing economy, the production of 'unimagined communities' and 'spatial amnesia' in the name of development has been rife since the onset of a planning promising all kinds of benefits not only to the nation–state but its people. What it has also delivered is a large-scale displacement of communities, physically ejecting them from the environments which they dwell in. In this mode of invisibilisation of peoples from their environments, under the auspices of nation–state policies and corporate interests, Nixon writes: 'The result is what I have called spatial amnesia, as communities, under the banner of development, are physically unsettled and imaginatively removed, evacuated from place and time and thus uncoupled from the idea of both a national future and national memory'

(Nixon 2011, 150). The dispossession of communities from their rightful resources is the primary focus of Nixon's urgency.

Fiction such as the ones analysed here, works to make visible the multifarious ways in which the human and non-human are placed in conflict over access to environmental resources. Narratives are able to imaginatively evoke and describe issues that the sociological solidity of claims in critical articles struggle with, especially in terms of the 'politics of scale' (see Blakey, 2020). This happens in a context of changed politico-economic arrangements that are ideologically sold as logical and pragmatic, both in the Global South as well as in the North. In *The Great Derangement* (2016), Amitav Ghosh's non-fiction book about global warming, the text communicates the process whereby his creative writing itself was spawned by a rising awareness of climatological and geological change that he could see before his eyes:

> I happened then to be writing about the Sundarbans, the great mangrove forests of the Bengal Delta, where the flow of water and silt is such that geological processes that usually unfold in deep time appear to occur at a speed where they can be followed from week to week and month to month. [,..] Portents of accumulative and irreversible change could also be seen, in receding shorelines and a steady intrusion of salt water on lands that had previously been cultivated. This a landscape so dynamic that its very changeability leads to innumerable moments of recognition.
>
> (Ghosh 2016, 5–6)

The dynamic ecology, that is described here, of the Sundarbans (in today's Bengal) suffusing the lives of people is also mirrored in the challenges of the earth in every part of world if we can or choose to see it. Representing anthropogenic climate change is another challenge as is the slow, incremental violence of climate change. The crises around water, in contrast, appears not so abstract or imperceptible in sites other than the Global North.

The great Himalayan watershed

The region encompassing the Hindu Kush Himalayas mountain range and the Tibetan Plateau (on the Chinese side) is now broadly known as the Third Pole. Its ice fields enclose the largest reserve of freshwater outside the two Polar Regions (see Figure 4.1). The significance of the region is apparent in that it is the source of the ten major river systems. The system makes possible irrigation, power and drinking water to almost two billion people in Asia. This constitutes over 24% of the world's population.

Rethinking the Himalayas and the Third Pole in the age of the Anthropocene is to tentatively look at some problems in evidence there in the area, problems related to environmental transition on a scale unknown to humans. It is also to pose questions on many different levels.[6]

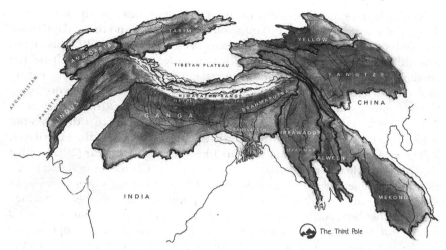

Figure 4.1 Third Pole span of the Himalaya freshwaters.

Source: Joydeep Gupta of https://www.thethirdpole.net/en/about/

At least a third of the huge ice fields in Asia's towering mountain chain are doomed to melt due to climate change, according to a landmark report, with serious consequences for almost two billion people. The 2016 report *The Hindu Kush Himalaya assessment: mountains, climate change, sustainability and people* is dire in its warnings. *The Guardian* reports: 'This is the climate crisis you haven't heard of', said Philippus Wester of the International Centre for Integrated Mountain Development (Icimod), who led the report. 'In the best of possible worlds, if we get really ambitious [in tackling climate change], even then we will lose one-third of the glaciers and be in trouble. That for us was the shocking finding'.[7] It is a detailed report with much to chew upon, but even if carbon emissions are dramatically and rapidly cut and succeed in limiting global warming to 1.5°C, 36% of the glaciers along in the Hindu Kush and Himalaya range will have gone by 2100. If emissions are not cut, the loss soars to two-thirds, the report found.

The glaciers are a critical water store for the 250 million people who live in the Hindu Kush-Himalaya region, and 1.65 billion people rely on the great rivers that flow from the peaks into India, Pakistan, China, and other nations. Himalayan peaks are said to be warming between 0.3°C and 0.7°C faster than the global average, the report shows. The meltings in the Himalayas are seen as corroborating the cascading effects of climate change not only upon these mountains, but water, biodiversity, and livelihoods as well.

The meltings in the Himalayas are seen as corroborating the cascading effects of climate change not only upon these mountains, but water, biodiversity, and livelihoods as well (Bolch et al. 2012; Xu Jianchu et al. 2009). Despite the scale of the crisis, it continues to be further aggravated by a sense of governmental inertia.

Research has tended to focus on the contemporaneity of the immediate disaster: the impact of glacial changes on local societies and their adaptations, eco-systems, and disaster coping mechanisms (Nüsser et al. 2019; Nüsser and Schmidt 2017). However, I argue, that in order to fully comprehend the current situation and overcome this inertia, a rethinking of the very nature of the question is necessary. Although scholars have used frameworks such water regime, water sites, and cryosphere (Moore et al. 2018; Nüsser et al. 2019; Orlove and Caton 2010; Scarborough 2003), I argue that this does not adequately capture the dimension of the question of water in the Himalayas.

The concern among certain scholars like William Goodyear (2012) is that the risky gamble taken by leaders in South Asia and China lies in the assumption that sufficient clean water will be available—from the Great Himalayan Watershed—such that the economic and political strategies being pursued through the last century are not at odds with the present, and the risk of future instability and war is low or non-existent.[8] Yet these assumptions are challenged: one result of this, the warning goes, is that some will have too little water while others like large swathes of Bangladesh will drown in too much of it. Kenneth Pomeranz writes that conflicts over water are everywhere but '[n]owhere are the stakes higher than in the Himalayas and on the Tibetan plateau: here the water-related dreams and fears of half the human race come together' (Pomeranz 2009, 1).

The great Himalayan watershed's system of precipitation, monsoons, glaciers, and water-cycle emanating from the Third Pole is under great threat. If things go unchecked, chances are that we would have not only initiated a great population transfer, refugee crisis, a war, ... but also the permanent destruction of a system that has sustained life and myriad cultures for millennia.

To talk about water or waterscapes is not only water as an essential resource and source of concern both for hydrogeologists and ecologists, and water drinkers in the Himalayas and those living around it. In this world 'nothing is supple and weak in relation to water. Yet of those things which attack the firm and unyielding, nothing is able to do better'. Laozi's words in chapter 78 of *Dao De Jing* from the 6th century B.C.E. resonate today in 'water-related dreams and fears'—both supple and unyielding at the same time—criss-crossing and interweaving the Himalayas and the Tibetan plateau for almost half the world's population.

In the early 1990s, rural people both in Sikkim in India and adjoining Tibet Autonomous Region were highlighting radical alterations in their environment and in their weather and seasonal patterns long before the mantra of 'climate change' (Tibetan *namshi gyurba*) had become incantatory and popular subject. People living in those regions reported greater thermal excursion between day and night, and unparalleled heat waves became later progressively inscribed in narratives of 'climate change'.

In thinking of the Anthropocene in the Himalayas, the ontology of water cannot be unassumingly simple or clear-cut in that a down-to-earth division and border between nature and culture is not tenable as Latour would point out. Karine Gagné writing for *Advanced Science News* has isolated three standpoints in Antropocenic terms: 'how water is produced as people interact with a sacred geography, how snowy peaks are produced as objects of morality through affective attachment and encounters, and how water is produced as part of multi-species assemblages, a process that has implications for the current changes in the climate, the weather, and the environment' (Gagné 2020, 4–5). The question of human moral attitude in its intense historical connection of give and take in the Himalayan landscape and its nonhuman things and objects, while determining the meanings invested and excavated by people living there, also paves the way forward for an understanding of water a special object.

Water as hyper

Jamie Linton in his book *What is Water? The History of a Modern Abstraction* (2010) makes us aware of the history of water as an abstract concept, stripped of its environmental, social, and cultural contexts. Reduced to a scientific abstraction—to mere $H[2]O$—this concept has given modern society licence to dam, divert, and manipulate water with apparent impunity. Part of the solution to the water crisis, for Linton, involves reinvesting water with social content, thus altering the way we see water.

In contrast to the hydrologic cycle, the hydro social cycle corresponds well, he argues, to emerging ideas and practices of water governance. It may be observed that the state-hydraulic paradigm of 'water management' is giving way to new modes of 'water governance'. The hydrologic cycle (see ICIMOD)[9] works well as a way of representing the nature of water for purposes of water management. This is because the hydrologic cycle conceptually isolates hydrological processes from ecological and social processes, thus making it possible to conceive of manipulating, or managing water as a discrete activity.

Reviewing the modern constitution of water, Linton's position is how it is not water itself, but its modern conception, which is in crisis. Discourse on water scarcity, hydrological stress and predicaments, the global water crisis, etc., assume a neo-Malthusian mode whereby 'quantification of abstract water [...is related] with the abstract quantification of people' (2010, 210).

Linton offers a perspective out of the impasse that our conception of modern water has led us to. *Hydrolectics,* as he calls it, foregrounds ideas of the hydrosocial cycle positing that 'the very nature of the circulation of water on earth, [...], has to be described in *social* as well as *hydrological* terms' (Linton 2010, 229; emphases in original). Linton's call is to insert ourselves in a personal bottom–up combat with decision-making processes. Importantly, also to become part of the solution by modifying the meaning of water (Linton 2010, 235). What he seeks is an obligation to 'ground' the pluralities of water.

In order to solve the impending water crisis, Linton suggests, we need to rethink how we see water and rediscover earlier social conceptions of water. Although Linton is in part Latourian (if 'part-ly' is possible), he does not consider water as part of the Anthropocene. If we consider water, for a moment, as nonhuman, then what is our 'human' relationship to or role in it? 'Earlier social conceptions of water' may not always be congruent with or entertained by those who want to emphasise the Anthropocene. The Anthropos of the Anthropocene reconfigures the human as a nonhuman geological force (Chakrabarty 2012, 15), signalling the collapse of the distinction of human and natural history (Chakrabarty 2012, 10). Human time-scales in the Anthropocene are de-prioritised, and we are required to think about deep or geological time. We are asked to contemplate the vast and less-human time scales and subjects. Water when tied to the Anthropocene and the hydro-social cycle, inevitably lead us to a consideration of power relations between humans themselves and non-humans. The dynamic range of the hydro-social cycle occurs within political and socioeconomic structures that are historically mediated; these relations in many ways outline the Anthropocene or the Capitalocene in their complex interconnections (see Parenti 2016).[10]

To think of water in this cosmology is also to be motivated by Timothy Morton's idea of the 'radical intimacy' of all things. 'The ecological thought is the thinking of interconnectedness.... It is a vast, sprawling mesh of interconnection without a definite center or edge. It is radical intimacy, coexistence with other beings, sentient and otherwise—and how can we so clearly tell the difference?' (Morton 2010, 7–8). In advocating for a profound relationality and interconnectedness of all things, Jane Bennett (2010) too, for example, places the human body (in its particularity) within a broader network or assemblage of other 'vibrant' things. She celebrates the 'nonhuman vitalities actively at work around and within us' (Bennett 2012, 231), and in this way she simultaneously challenges the stability of the human as a bounded and coherent entity, and the notion of the passivity of nonhuman matter. Water, I am evidently assuming, would be a salient part of this matter.

If that is the case, for Morton the most interesting questions then become: how do I get to coexist with them, with these nonhuman entities? To what extent? In what mode or modes? (or how do we delineate the water-packed Himalayas or the monsoons?). The role of water as a symbol, metaphor, or discursive element in its relationality is vivid as my reading of the novels indicated.

However difficult, is it possible to imagine water from *its own* hydrocentric perspective? Paresh's struggle, in Joshi's novel, with coming up with an adequate description of water for his daughter can be typified as a problem of perspective. If you, as an addressee, can (hear what I am saying now, or read what I am writing now) you are most likely a human being and you can never sense what it is like to be water (although you consist of roughly two-thirds water). It is because the real water of your body is far removed, 'withdrawn', as Morton would have it, even from yourself.

Water is a typical *hyperobject* in that sense, existing on multiple scales, from tiny molecules to the vast hydrological cycle the role of water as a symbol, metaphor, or discursive element is vivid.[11] Although Morton's idea of the *hyperobject* has been criticised, I argue that it makes sense to consider water as such. Eco-critic Ursule Heise has issued a demurral that in Morton's articulation, everything tends to being a *hyperobject,* rendering the concept somewhat vacuous. Heise argues that Morton makes 'so many self-cancelling claims about hyperobjects that coherent argument vanishes like the octopi that disappear in several chapters in their clouds of ink, Morton's favorite metaphor for the withdrawal of objects from the grasp of human knowledge' (Heise 2014). Water does withdraw, does slip away, but in its easy access and routineness, and its abstractness if you like, as well as its taken-for-grantedness, allows it, my argument is, to be treated productively as a *hyperobject* as I show below.

Water does not easily cease to exist. Ice can melt, liquid water can freeze, boil, evaporate, percolate, etc. but it will remain as water as long as the H[2]O molecule does not break up into hydrogen and oxygen. Spatial scale is important since it affects how water behaves in relation to other objects. The H[2]O molecule has characteristics that make it different from the ocean because trillions of H[2]O molecules behave differently than one molecule does due to gravity, winds, salinity, water temperature, density, and all the rest.

Even if a substance ('water') is made up of other substances ('hydrogen' and 'oxygen') it is still one substance different in kind from its parts. Substances are therefore non-dialectical; there is nothing contrary to them. There is no 'anti-water'. If contraries exist it is only among qualities (such as steam and ice) (Bryant 2011, 73).

A substance can actualize different qualities at different times and it can also fail to do so (Bryant 2011, 85). Statistically normal amounts of rain in a given waterscape are good for crops, but too low amounts and we have a drought and too much water we have a flood. These differences actualize different qualities and capacities of rainwater (such as people dying of thirst, sustaining plants, transporting contagions, and drowning people).

What scale should one choose while studying water? For an object oriented theorist like Morton (2014), his concept of the *hyperobject* offers some solutions, one can argue, when addressing water as a special kind of object. *Hyperobjecs* are massively distributed objects, and humans and other objects are located within them often without being able to perceive them directly. Water is one such object. No doubt, it adheres to you, it touches you, and you see what is our taken-for-grantedness of it.

Morton puts forward the idea of *hyperobjects* in order to rethink what our necessarily entangled relationships are to things. He argues that that *hyperobjects* are viscous—they adhere to you no matter how hard to try to pull away, rendering ironic distance obsolete. He argues too that they are also nonlocal. That is, *hyperobjects* are massively distributed in time and space such that any particular (local) manifestation never reveals the totality of the *hyperobject*.

When people feel raindrops falling on their heads, they are experiencing climate, in some sense. In particular they are experiencing the climate change that is described as global warming. But they are never directly experiencing global warming as such. Nowhere in the long list of catastrophic weather events—which will increase as global warming takes off—will one find global warming.

But global warming, as we know scientifically, is as real as this sentence.[12] Not only that, it is viscous, Morton posits. It never stops sticking to a person, no matter where one moves on the planet Earth. How can we account for this? By arguing that global warming, like all *hyperobjects*, is nonlocal: it is, in other words, massively distributed in time and space.

What are the implications of this? It suggests that any experience of the weather in the here and now is a false immediacy. It can never be the case that those raindrops only fall on my head. They are always a manifestation of global warming in that sense. In an age of ecological emergency—in an age in which *hyperobjects* start to oppress us with their terrifying strangeness—the point to take home is that humans have to become habituated to the fact that locality, the definite sense of emplacement (that even displaced refugees in a detention centre have), is always a false immediacy.

There are certain characteristics that Morton (2014) delineates in his idea of *hyperobjects*. While they have been alluded to above, a recapitulation is helpful:

1 They are *viscous*. This means that they stick to other objects involved with them because the *hyperobject* pre-exists many other objects and incorporates them. The hydrological cycle pre-exists all humans, and since we constantly interact with water we are stuck to it.

2 *Hyperobjects* are *molten*. This means that they contradict the Newtonian idea of a fixed, concrete, and consistent space-time. They exist on a scale beyond humans. This could be the Indian Ocean or water on other planets.

3 Since they are widely distributed they can never be observed in a specific local manifestation. This *non-locality* means that information of the object is distributed among parts that occupy a seemingly non-continuous 'space-time'. Water in my body is seemingly cut off from the glaciers in the Himalayas, yet they are clearly local manifestations of the same hydrological cycle.

4 *Hyperobjects* are *phased*. A phase-space is the set of all possible states of a system/object. When the time that one object emits intersects with the time of another object we get an interference pattern (phasing). The Earth's orbit around the sun affects seasonal changes in the hydrological cycle and the moon's orbit around the Earth affects tides at the same time. Surely humans experience tsunamis but those are simply 'snapshots' of the *hyperobject* that occupies a more complex space (Morton 2014, 70). Selecting one 'thin slice' of the *hyperobject* to analyse is just as likely to weaken our understanding of the *hyperobject* as it is to improve it (2014, 70). Morton uses the study of an iceberg (pictured on the front cover of his book) to illuminate phasing. Given adequate light and distance, humans can see the tip of an iceberg. However, 90% of the iceberg

is under water. The scientific reaction to this dilemma is to move the camera underwater, but this choice distorts the part of the iceberg that sits above water, rendering a complete picture impossible. The scientist faces other choices as well. Should she move in close to the iceberg, or is there some perfect critical distance? If she gets close she may miss important context, but if she moves too far away, she misses the undulations and imperfections of the surface. Should she drill or melt part of the iceberg? If she did, she might end up knowing more about the iceberg she intended to observe, but she would have fundamentally altered the object of study in the process. Should she map the movement of the iceberg? This may tell her something about the pedigree of the object, but she can't trace back forever, and she will never perfectly know the future itinerary of the iceberg. (None of these dilemmas deny the reality of the iceberg; McDonald's novel, in some ways tries to make sense of the trajectory and nature of the iceberg from the Antarctic being towed to the Bay of Bengal.)

5   The effects of *hyperobjects* are shown *interobjectively*. Interobjectivity is the 'abyss in front of things', it is what we usually term 'space-time'. They are created by the exo-relations of many objects. An object can only perceive a *hyperobject* upon another object. They are interobjective because they constitute the mesh that floats in and around other objects (83). *Hyperobjects* are causal webs that leave footprints on other objects. In other words, many *hyperobjects* are 'crisscrossing' force fields (93). Global warming leaves a footprint on the back of our necks.

If we consider the *hyperobject par excellence*, that is, climate change, what can we say? Climate change is not like the sublime mountain, Everest or Kanchenjunga (take your pick of any high peak) that, in finite time, gives us an intimation of an eternal, infinite, and unchanging nature. Climate change is given to us always as a thoroughly regular event: rain, cyclones, heat waves, typhoons, and even rising sea levels (because considering a long enough time frame, the world has gone through periods of warming, cooling, and beyond that periods before life and before earth). We experience the *hyperobject* in fragments that add up to a finite 'whole' but in a finitude so radical that unlike the infinite it is not given as some eternal horizon, but just as a volatility that we experience only by way of things that 'we' have always known, but just not in this rhythm.

Conclusions: work in progress

There can be no conclusion to such a mammoth issue that is raised in the chapter. Whatever I have written in the sections above is not even the tip of the proverbial iceberg. There is clearly much heft to the postcolonial argument that Agarwal and Narain make regarding environmental colonialism, which continues apace in different parts of an uneven globe; there is too, beyond that, the larger geological time framework, à la Chakrabarty and others, that delineates how we have all been placed in the same fragile boat and need to address questions collectively. Resourceful thinking while addressing the Himalayan water problems, both in the present and the future, can take different forms. For instance, resonating local traditions can be tremendously useful in producing a rejuvenated discourse as well as practice(s) in water thinking and usage. The theoretical concept *qi* 氣 takes water vapour as its model but it also extends to suggest water in its other forms, from hardened ice, to moving water, to dissipating vapour (Allan 2003). Somewhat similar is the Tibetan notion of *namshi gyurba* (Diemberger 2012). A reading and critical analyses of such philosophical-spiritual resources provide a way in which water becomes meaningful to people directly affected by the crisis. Something that is beyond the scope of this short chapter but important to flag are: State policies (Atteridge et al. 2012; Xu and Grumbine 2014), scientific and academic literature to be found in a plethora of works (Arupjyoti 2019; Baghel 2012; Bolch et al. 2012; Ghosh 2016; Saikia 2019; Strange 2016; Wester et al. 2019; Xu Jianchu et al. 2009; Xu et al. 2009), and philosophical discourses (Yao Zhihua 2014) produced in and on the region. They can in their own ways provide a sense of the colossal crisis in which the great Himalayan watershed's system of precipitation, monsoons, riverine systems, and glaciers finds itself.

Notes

1. Original in 'Tai Yi, shui yu Zhongguo gudai yuzhou sheng cheng lun', The text, scholars agree, has an unknown date. Yao's source is a Song Dynasty citation in the *Taixuanbu* of the *Dàozàng* or the Taoist Canon. The line quoted is found in current versions of the text. (Allan 2003, 276).

2. 'Himalayas' in this essay represents the area encompassing the large mountain range of Asia, that runs from Afghanistan through Pakistan and India/Nepal to China along with the Tibetan Plateau and the Indian subcontinent. The mountain range separates these countries/regions, which depend on water flowing from the Himalayas.

3. Chakrabarty has been criticised, among other things, for his contention that humanity as a whole is responsible for climate change. This implies that humanity as a whole, and not just those who caused it primarily, has a collective responsibility to tackle the problem (2012). Boscov-Ellen has argued that reconstructing 'the intellectual-historical arc that lead Chakrabarty to [the] surprising shift can help to soften antagonisms between Marxism and postcolonial thought that currently stand in the way of an adequate understanding of the climate problem' (2020, 70).

4. The current consensual narrative is that humans as we move linearly into the future are exceptionally disruptive. Dorian Sagan holds that a better metaphor is that we are on a kind of 'Möbius trip'. 'Those on a Möbius strip walk a cryptically twisted path. The Möbius strip seems a straight line but returns its voyagers to their starting point. Evolving life, especially in its Nietzschean ecology mode, may be similar' (2016).

5. One of the consequences arising from the Anthropocene is the idea that the human-nature dualism is untenable. Nature stands hybrdised in that it is part of the various processes that connect society. Hybrids are nature-culture outcomes.

These hybridisation processes allow for multiple agencies to be distributed in networks, including non-human entities and processes (see Latour 2004).

6. Not a very productive strategem to do this is to appropriate the idea and reality of the Third Pole and render it in terms of a Sinocentism where China is launched as a Third Pole culture that is ostensibly post-imperial and post-colonial. For a critique, see Poddar and Zhang (2020).

7. 'A third of Himalayan ice cap doomed, finds report', (Carrington 2019).

8. The Great Himalayan Watershed refers to the hydrological phenomenon that has created richly diverse ecosystems and the right conditions for some of the world's earliest agricultural and urban centres. Ruth Gamble calls it 'The unwinnable contest for Himalayan water resources' (2019).

9. The International Centre for Integrated Mountain Development (ICIMOD) is an intergovernmental knowledge and learning centre working on behalf of the people of the Hindu Kush Himalaya. See https://www.icimod.org/

10. Capitalocene has been pitted against the idea of the Anthropocene. In an interview, Jason Moore says: 'Capitalocene is a kind of critical provocation to this sensibility of the Anthropocene, which is: We have met the enemy and he is us. So the idea that we're all going to cover our footprints, we're going to be more sustainable consumers, we're going to pay attention to population, are really consequences of a highly unequal system of power and wealth'. *Wired*, 09.20.2019. https://www.wired.com/story/capitalocene/

11. The suffix 'object' is used here in the sense that object-oriented ontology **posits** objects by **replacing** the real with these posited objects; thinking with objects is not the same as thinking with facts (as in Ludwig Wittgenstein), thinking with events (as in Giles Deleuze), thinking with Truths (as in Alan Badiou), or thinking with Subject (Slavoj Zizek). Subjects in the plural and in small 's', of course, think with or against objects.

12. Even a popular magazine like *National Geographic* in its issue of January 31, 2019 asks the question 'Is global warming real?' and answers that the scientific consensus is overwhelming that our planet is getting warmer.

References

Allan, Sarah. 2003. "The Great One, Water, and the Laozi: New Light from Guodian." *T'oung Pao* 89, no. 4/5: 237–285.

Agarwal, Anil, Sunita Narain, and Centre for Science and Environment (New Delhi, India). 1991. *Global warming in an unequal world: a case of environmental colonialism.* New Delhi, India: Centre for Science and Environment.

Atteridge, Aaron, Manish Kumar Shrivastava, Neha Pahuja, and Himani Upadhyay. 2012. "Climate Policy in India: What Shapes International, National and State Policy?" *AMBIO* 41, no. 1: 68–77.

Baghel, Ravi. 2012. "Knowledge, Power and the Environment: Epistemologies of the Anthropocene." *Transcience: A Journal of Global Studies* 3: 1–6.

Bennett, Jane. 2010. *Vibrant Matter: A Political Ecology of Things.* Durham: Duke University Press.

Bennett, Jane. 2012. "Systems and Things: A Response to Graham Harman and Timothy Morton." *New Literary History* 43, no. 2: 225–233.

Bishta, Minbahadur. 1991. "Thus a Nation Pretends to Live." In *Himalayan Voices: An Introduction to Modern Nepali Literature*, edited and translated by Michael Hutt. Berkeley and L.A.: University of California Press.

Blakey, Joe. 2020. "The Politics of Scale through Rancière." *Progress in Human Geography.*

Bolch, T., A. Kulkarni, A. Kääb, C. Huggel, F. Paul, J.G. Cogley, H. Frey, et al. 2012. "The State and Fate of Himalayan Glaciers." *Science* 336, no. 6079: 310–14.

Boscov-Ellen, Dan. 2020. "Whose Universalism? Dipesh Chakrabarty and the Anthropocene." *Capitalism Nature Socialism* 31, no. 1: 70–83.

Bryant, Levi R. 2011. *The Democracy of Objects.* Ann Arbor: Open Humanities Press.

Carrington, Damian. 2019. "A Third of Himalayan Ice Cap Doomed, Finds Report." *The Guardian*, 4 February 2019. https://www.theguardian.com/environment/2019/feb/04/a-third-of-himalayan-ice-cap-doomed-finds-shocking-report.

Chakrabarty, Dipesh. 2012. "Postcolonial Studies and the Challenge of Climate Change." *New Literary History* 43, no. 1: 1–18.

Diemberger, Hildegard. 2012. "Deciding the Future in the Land of Snow: Tibet as an Arena for Conflicting Forms of Knowledge and Policy." In *The Social Life of Climate Change Models: Anticipating Nature*, edited by Kirsten Hastrup and Michael Skrydstrup. New York and London: Routledge.

Gagné, Karine. 2020. "Water, Reciprocity, and the Anthropocene in the Himalayas". *Advanced Science News*, May 26, 2020. https://www.advancedsciencenews.com/water-reciprocity-and-the-anthropocene-in-the-himalayas/.

Gamble, Ruth. 2019. "The Unwinnable Contest for Himalayan Water Resources." *East Aisa Forum*, June 2, 2019. https://www.eastasiaforum.org/2019/07/02/the-unwinnable-contest-for-himalayan-water-resources/.

Ghosh, Amitav. 2016. *The Great Derangement: Climate Change and the Unthinkable.* Chicago: University of Chicago Press.

Goodyear, William. 2012. "The Water Security Gamble in the Greater Himalayan Watershed." October 2, 2012. http://southasiajournal.net/the-water-security-gamble-in-the-greater-himalayan-watershed/.

Heise, Ursula K. 2014. "Ursula K. Heise Reviews Timothy Morton." *Critical Inquiry*, June 4, 2014. http://criticalinquiry.uchicago.edu/ursula_k._heise_reviews_timothy_morton.

Joshi, Ruchir. 2001. *The Last Jet Engine Laugh.* London: Flamingo.

Latour, B. 2004. *Politics of Nature. How to Bring the Sciences into Democracy.* Cambridge: Harvard University Press.

Linton, Jamie. 2010. *What Is Water? The History of a Modern Abstraction.* Vancouver, Toronto: UBC Press.

Mcdonald, Ian. 2004. *River of Gods.* New York: Simon & Schuster.

Moore, Michele-Lee, Karena Shaw, and Heather Castleden. 2018. "'We need more data!' The politics of scientific information for water governance in the context of hydraulic fracturing." *Water Alternatives* 11 no. 1: 142–162.

Morton, Timothy. 2010. *The Ecological Thought.* Cambridge, Mass: Harvard University Press.

Morton, Timothy. 2014. *Hyperobjects: Philosophy and Ecology after the End of the World.* Minneapolis: University of Minnesota Press.

Morton, Timothy. 2016. *Dark Ecology: For a Logic of Future Coexistence.* New York: Columbia University Press.

Mueller, Max. 1897. *Sacred Books of the East.* Oxford: Clarendon Press.

Nixon, Rob. 2011. *Slow Violence and the Environmentalism of the Poor.* Cambridge, MA: Harvard University Press.

Nüsser, Marcus, Juliane Dame, Benjamin Kraus, and Ravi Baghel. 2019. "Socio-Hydrology of 'Artificial Glaciers' in Ladakh, India: Assessing Adaptive Strategies in a Changing Cryosphere." *Regional Environmental Change* 19, no. 5: 1327–1337.

Nüsser, Marcus, and Susanne Schmidt. 2017. "Nanga Parbat Revisited: Evolution and Dynamics of Sociohydrological Interactions in the Northwestern Himalaya." *Annals of the American Association of Geographers* 107, no. 2: 403–415.

Orlove, Ben and Steven C. Caton. 2010. "Water Sustainability: Anthropological Approaches and Prospects." *Annual Review of Anthropology* 39, no. 1: 401–415.

Parenti, Christian. 2016. "Environment-Making in the Capitalocene: Political Ecology of the State." In *Anthropocene or Capitalocene: Nature, History, and the Crisis of Capitalism*, edited by Jason W. Moore. Oakland, CA: PM Press.

Poddar, Prem, and Lisa Lindkvist Zhang. 2020. "China as 'Third Pole Culture':Between theorizing and thought work." In *Redefining Propaganda in Modern China: The Mao Era and Its Legacies*, edited by James Farley and Matthew D. Johnson. London and New York: Routledge.

Pomeranz, Kenneth. 2009. "The Great Himalayan Watershed: Water Shortages, Mega-Projects and Environmental Politics in China, India, and Southeast Asia." *The Asia-Pacific Journal* 7, no. 30: 1–29.

Sagan, Dorian. 2016. "Möbius Trip. The Technosphere and Our Science Fiction Reality." *Technosphere Magazine*, November 15, 2016. https://technosphere-magazine. hkw.de/p/Mobius-Trip-The-Technosphere-and-Our-Science-Fiction-Reality-fq6MUxZjiBx7pzKPMKZfcb.

Saikia, Arupjyoti. 2019. *The Unquiet River: A Biography of the Brahmaputra*. New Delhi: Oxford University Press.

Scarborough, Vernon. 2003. *The Flow of Power: Ancient Water Systems and Landscapes*. Santa Fe: School of American Research Press.

Strange, Veronica. 2016. "Infrastructural Relations: Water, Political Power and the Rise of a New Despotic Regime." *Water Alternative* 9, no. 2: 292–318.

Wester, Philippus, Arabinda Mishra, Aditi Mukherji, and Arun Bhakta Shrestha. 2019. *The Hindu Kush Himalaya Assessment: Mountains, Climate Change, Sustainability and People*. Cham: Springer.

Xu Jianchu, R. Edward Grumbine, Arun Shreshta, Mats Eriksson, Xuefei Yang, Yun Wang, and Andreas Wilkes. 2009. "The Melting Himalayas: Cascading Effects of Climate Change on Water, Biodiversity, and Livelihoods." *Conservation Biology* 3, no. 2: 520–530.

Xu, Jianchu, and R. Edward Grumbine. 2014. "Integrating local hybrid knowledge and state support for climate change adaptation in the Asian Highlands." *Climatic Change* 124, 93–104.

Yao Zhihua. 2014. "One, Water, and Cosmogony: Reflections on the Rgveda X.129 and the Taiyi Sheng Shui." In *Brahman and Dao: Comparative Studies of Indian and Chinese Philosophy and Religion*, edited by Yao Zhihua and Theodor Ithamar. Lanham, MD: Lexington Books.

5 Decolonising gender

Witches, nomads, and the colonial rule

Nazila Ghavami Kivi

Introduction

During winter 2020/2021, the Copenhagen Museum of Contemporary Art Kunsthal Charlottenborg ran an ambitious and thorough exhibition on the Nordic witch-hunts that took place during the sixteenth to eighteenth centuries (Karasyk and Ugelvig 2020a). The exhibition, Witch Hunt, to which I had the pleasure to contribute with an essay in the accompanying book (Karasyk and Ugelvig 2020b), bravely engaged with topics usually either forgotten (Stoler 2011) or glorified in European and Scandinavian collective memory (Nonbo Andersen 2013), like the colonial history of enslavement and looting that has led to the wealth of the region (Suárez-Krabbe 2017). Witch Hunt connects the witch-hunts to colonial history and racism, engaging art that addresses the Danish colonies in the Danish West Indies (currently US Virgin Islands) (Jensen 2018), Denmark's relationship to Greenland (Mortensen and Maegaard 2019) and other Nordic countries' past and present relationship with their indigenous peoples, the Sami (Dobbin 2013). Engaging with the witch and the history of the witch-hunts has increasingly become part of various feminist and queer (Lykke 2010; Valencia 2018) movements across the world (Opstrup 2018). The figure of the witch, as an underdog from the past who resurrects from the dead with a vengeance (Ghavami Kivi 2019a) seems like a natural object of identification for individuals and groups who are suppressed and marginalised due to their gender, race, sexual desires, and identity. However, the renewed and understandable interest in the witch as a feminist icon shares fate with numerous other feminist causes, as it relates to the question of the complex relationship between gender, race, and colonialism, and between European imperialism and the witch-hunts, which is often omitted. In many engagements with the witch, these connections seem forgotten. Even when there is thorough involvement with colonial history, the connection between the historical witch-hunts and contemporary European racism, especially anti-Muslim sentiments, seems to be absent. Critical reflections of the complexity between the European witch-hunts, colonial history, and orientalism are few and far between. In this chapter I attempt to counter this lack of reflection, particularly on the connection between race, gender,

DOI: 10.4324/9781003172413-6

and the witch. Furthermore, I will illustrate how orientalism, anti-Semitism, and the witch-hunts are deeply interconnected not only historically but in contemporary Western thinking as well. In order to truly understand how the witch is connected to gender, and gender to coloniality, I will start by delineating the most important categories regarding the witch-hunts: gender and The Woman. If one wants to examine how gender can be 'decolonised' one must begin by shedding light on how colonialism and the Western-imperialist notions of gender are two sides of the same coin (Lugones 2007). Only by a thorough understanding of these connections, one can begin to appreciate how The Witch, as a figure and an icon, can lead the way to decolonise gender.

In order to understand the importance of reproduction (the production and maintenance of life and the physical and emotional wellbeing of people, see Bhattacharya 2017; Federici 2014) in critical feminist and decolonial studies, it is important to understand the centrality of reproduction and reproductive politics in nation states. Reproduction and the right hereto is critical in nation states because it involves the control of borders (who belongs where), population management (who can reproduce and especially, who cannot), the right to free movement of people (migration policies) (Braidotti 1994), and capital (class disparities in reproduction, human capital as workers and as the enslavement of people for the production of capital). These connections all involve gender and hence are important to grasp before moving on to the next parts of this chapter.

It is from this point of departure that I will expand on the figure of the witch as crucial to subversive femininity. The witch opens up possibilities to dissect and queer the intersections of gender, race and coloniality, and to problematise how gender came to exist as a defining category in Western culture all together.

Along with a growing number of southern feminist scholars, this chapter contributes to increasing our understanding of how gender and race are co-constructed and co-dependent categories that infuse all modern Western thinking (Lugones 2007; Oyewumi 1997). Questioning such categories as facts is vital in order to decolonise the way we understand human life, and to work towards new liberatory ways to describe and comprehend being human beyond racial and gender hierarchies.

It is here that the witch lends us a helping hand to understand how the violent histories of gender, race, and death have been inextricably linked together in coloniality through centuries. With her resurrection from the dead as a Simorgh or Phoenix, The Witch will guide us to restart the revolutionary process of abolishing oppressive categories that bind us and limit the freedom of human (and more-than-human) existence.

First things first: gender and femininity as colonial problems

If you have read texts on gender and feminist studies before, it is likely that you have met the notion that there is 'sex' and there is 'gender' as if they were two different categories, the first describing biological bodily phenomena as a factual state and the latter, gender, as a social category. This dichotomy is, however, challenged by several critical gender scholars, among them feminist philosopher Donna Haraway (1990), Deborah Findlay (1993), and Nina Lykke (2010) who emphasise that even a biological 'fact' as sex is in itself subject to political and social conditions. Haraway introduces the concept of situated knowledge: that knowledge and the production thereof is always partial, subjective, power imbued, and contextual, even when presented as objective and factual (Haraway 1990). The acknowledgement that knowledge is situated in time, space, and context and that there is no objective and unbiased 'truth' waiting to be found, challenges the white, male supremacist idea that there is a god-like objective researcher without preconceptions of his own who produces neutral and unbiased knowledge, immune to power relations and to political, social and cultural influences. To imply that gender is social while sex is a biological fact is to overlook the history of sexuality in Western thought and how it is made into a defining category permeating all layers of social and political life. When the narrative about gender equality as a Western ideal and value emphasises progressions in women's rights (Sabsay 2012), LGBTQA+ rights and protection (Puar 2017), and hallmarks sexual liberty during the last forty to fifty years as proof of Western superiority to 'other' cultures (Sabsay 2016), one could turn the mirror and ask: Why didn't women have suffrage (the right to vote) to begin with in 'modern' society, and why was The Enlightenment so deeply infiltrated with homophobia and misogyny (von Eggers 2016)? The answer lies in how difference was regarded in Western thought and modern science. As an effect of modernity, we might all have been trained to think about difference as something that has to be ranked and organised, as something that has to be put in a systematised order; just think of Charles Darwin's theory of evolution (1859) that builds upon a strict hierarchical system in which nature can be understood as a linear progression from 'lower' to 'higher' forms of being; or the Swedish botanist and physician Carl von Linné who is also known as the father of biological racism, who ranked humans in different states of evolution and progression, with darker skin colours as the least progressed (1732). It is in the same categorisation of human beings that sex and gender became defining characteristics as did race. It has been documented that biological science and biological racism has often resembled (white) women to 'other' races, that is, non-White people, and has approximated 'feminine' acts such as breastfeeding to more animal-like activities hence emphasising the shared features of women and animals (Schiebinger 1993). This also implies that as 'animality' is part of the inferiorisation of other humans, animals themselves are also inferiorised (Jackson 2020). The distinction between nature and culture in western thought is indeed also a hierarchical distinction. Based on this, it is safe to claim that much Western thought regards difference in a hierarchical way rather than as an organic diversity of being. The idea of gender as defining one's existence is a clear demonstration of this tendency, where a strict order with male coded individuals as superior to female coded ones has prevailed for centuries. In order to demonstrate how this has been in play in critical issues regarding human life and reproduction, let me introduce some examples of how western modern science, especially medicine and the science about women, have influenced questions as defining as the right versus obligation to reproduce, the definition of normalcy and insanity and the human cost of the science of gynaecology.

The weeping uterus

Philosopher and feminist scholar Deborah Findlay (1993) completed an anthropological study of Canadian gynaecology textbooks from the 1950s and found that sociocultural norms affect the production of knowledge about women's bodies, yet this knowledge is presented as if existing outside of social contexts, hence decontextualised and offered as objective facts. Findlay problematises the habitual splitting of social knowledge and biomedical knowledge while she emphasises that biomedical knowledge *is* in fact social knowledge. She exemplifies this by documenting how doctors and gynaecologists used 'facts' about the female body as a means to dictate normalcy and acceptable femininity, hereby outlining what kinds of womanhood and femininity were unacceptable. In this way, knowledge about women was used as a means to control women, their bodies and their actions. In the 1950s especially, many doctors and gynaecologists in the United States (US) and Canada were politically conservative and saw women's entry into the job market as a threat against the social order. For these doctors, women's natural place was at home, giving birth to as many children as possible. This pronatalist (pro-birth) approach was believed to be a guard against a wide number of phenomena such as unemployment and immigration of non-White workers from the Global South and communism (seen as a global threat to capitalism at the time), thus making white, middleclass-women responsible for the stability and 'pureness' of society. These ideas were presented as biomedical knowledge, for instance, one of the authors of the gynaecology textbooks that Findlay explored illustrated this by using the case of a woman in her twenties who had been pregnant non-stop since the age of fourteen and argued that she was the perfect woman, health wise, since she had never menstruated. He further described menstruation as 'the weeping of a disappointed uterus, sad for not fulfilling its true purpose' (Findlay 1993; Ghavami Kivi 2015a). The pathologisation of an otherwise natural phenomenon like menstruation and thereby of women not being pregnant aided in defining the role of women as potential mothers and housewives, limiting their participation in societal matters in the public space.

To put this in further perspective it is useful to have in mind that during the 1940s, under World War II, it was mostly women who were upholding production in society, when the United States eventually became involved in the global war, both regarding war machinery and everyday food and goods production, while many men were stationed as soldiers in Europe. With the return of men from war, jobs were in high demand again and women once again seen as threats to men's occupational options. Going a few decades back in time to 1920s and 1930s, a similar pattern and reasoning is evident in the work of US physician Robert Frank (Taylor 2006), who was one of the first to problematise and write about what later came to be known as premenstrual syndrome (PMS), PMS (Ussher 2006). Frank in his time called it premenstrual tension and believed that joining the work force and entering the public space in general caused tension and disturbance in women. As a curious fact, Frank was obsessed with ovaries, he believed that they made women unstable and irrational and hence incapable of making decisions at work. He also believed that women's natural place was at home, bringing up children. Similar to the 1950s when there was a lack of jobs for men returning from war, the Wall Street crash in 1929 led to the Great Depression and made jobs scarce during the 1930s. Women were seen as competitors to men. This entailed weaponising menstruation, reproduction, and women's bodies in the political movement against women's freedom. Even global politics regarding migration and the free movement of people and where they (could migrate to) work became part of the problematisation of menstruation. As it will appear below, making the private political, as one of the feminist movement's oldest and most durable slogans goes, is a literal matter when looking at the attempts to regulate the feminine body and ideals of womanhood in the name of modern medicine and gynaecology.

The fragile woman and the stationary mother

As mentioned above, Robert Frank was preoccupied with ovaries and the 'problem' hormone, oestrogen. This hormone is until this day still under scrutiny for causing women various conditions from depression to PMS (Eriksson et al. 2008), even without any biomedical-model evidence so far (Figert 2005). Considering oestrogen's history in Western medical and social history sheds more light on the interconnectedness of the roots of sex and gender oppression in European thought. Professor of Science and Technology Studies Nelly Oudshoorn (1994) has uncovered the social construction of gender and the female body in Western reproductive medicine at its roots. The claim that women are secondary to men was reflected in western medical science's discovery of the sex hormones oestrogen and testosterone that took place in the sexually restrictive Victorian era (Oudshoorn 1994; Roberts 2002). By these discoveries, the perception of gender shifted from the one-sex model to the two-sex model (Soble 2003). The one-sex model viewed the white man as the only normal human and all others as anomalies, while the two-sex model viewed white women and men as opposites with a distinct hierarchy that placed the man on top of the rank order (Soble 2003). The female hormones, therefore, were seen as inferior and ascribed properties that fit the ideals of European middleclass womanhood: weakness, sexual passivity, imbued with maternal instincts and stationed at home, where they would keep themselves busy with arts and crafts. Anthropologist Emily Martin (1996) has made a similar study of the way reproduction is discussed in medical terms (by doctors and subsequently by the general public) in US culture in the 21st century where male reproductive functions are described in heroic and grandiose terms while women's reproduction is presented as the opposite.

As a testimony of how this belief was ideological, selective and not a matter of 'fact' about *all* women, it is interesting to observe that these feminine ideals of stillness, fragility, and motherhood as sacred were not expected from working class or enslaved women, who worked hard side by side with men and weren't allowed the 'luxury' of confined womanhood. Angela Davis (1981) has documented how the legendary abolitionist Sojourner Truth's famous speech 'Ain't I A Woman?' at the Women's Rights Convention in Old Stone Church, Ohio, in 1851 reflected exactly this difference in how women were perceived depending on their class affiliation. Sojourner Truth proclaimed:

> I have ploughed and planted, and gathered into barns, and no man could head me! And ain't I a woman? I could work as much and eat as much as a man - when I could get it - and bear the lash as well! And ain't I a woman? I have borne thirteen children, and seen most all sold off to slavery, and when I cried out with my mother's grief, none but Jesus heard me! And ain't I a woman? (61–62)

In further emphasising the point Davis (1981) quotes Karl Marx, to show how women in England were used to haul canal boats instead of horses because women were cheaper to 'maintain' than horses or any machinery. Following this argument, she points out that during slavery in the United States, enslaved women were in higher demand than men in the Southern states. No fragility or softness was afforded these women when it was convenient to hold them to as hard working standards as men. The idea of women as weak and fragile is hence only an ideological conveniency to keep them out of the work force and out of political influence.

Gynaecology and slavery

One of the most cruel and horrifying examples of how modern medicine, slavery, and gendered violence intersect stems from the 'father' of modern gynaecology US physician and slave owner J. Marion Sims who conducted numerous experiments on enslaved Black women between 1845 and 1849 (Wall 2006). As Elise Thorburn (2017) documents:

> Sims' career was built upon the repair of vesicovaginal fistulas, and the techniques for such repairs were honed over four years (1845–1849) of experimentation in which he performed dozens of unanesthetised surgeries on three enslaved black women—Anarcha, Betsy, and Lucy. (161)

These and other women were properties of Sims and he conducted his experiments not only for the cause of medical development but for capital as well. Enslaved women, who were seen as 'breeders', were in high demand due to their productive and reproductive capacities, as they would give birth to children who would be enslaved from birth and add to the slave-owners' capital (Thorburn 2017). Many enslaved women were too young to be bearing children, but their children were an income source for slave owners so they were not spared numerous childbirths from an early age. Giving birth while not yet fully mature often resulted in a condition called vesico-vaginal fistula, a leak between the bladder or the colon and the vagina, causing pain, and socially stigmatising incontinence. Sims focused his efforts on finding a cure for this condition and conducted numerous non-consented experiments on enslaved women, exposing them to pain and suffering. Not until he had developed a successful method did he start curing white middle-class women for the same condition, this time using pain relief that had not been afforded to his black victims.

By exploring other gynaecological landmarks during this time, it becomes evident how the 'discovery' of the feminine body resembles the discovery of land (Kaminsky 2018) with many reproductive body parts named after the male doctor who first described them in a textbook, the most famous being the Fallopian tubes (the uterine tubes) and Bartholin's glands (lubricatory glands in the vulva). This aligns with the idea of women as (wild) nature while men were seen as the creators and bearers of culture and civilisation (Rosaldo and Ramphere 1985), while simultaneously creating a hierarchy between women depending on race and racialisation.

The above examples are just a few cross-sections in the history of 'the science about women' that one could call gynaecology. Of importance is the fact that these examples are not reserved for stories from the past but are ongoing and continuously take place in contemporary times (Ghavami Kivi 2019b). Whil the claims of 'biological racism', where racism is justified in biological terms, have been found to be obsolete and pseudoscientific, the beliefs that underpin such claims are deeply ingrained in medical history, such that ascribing them to the past avoids responsibility while the disparity that stems from racist ideologies is still in effect (Schulz and Leith 2006). Medical history and especially the science about women is still based on misogyny in our present time. Women are still believed less when they have pain (Riva et al. 2011). In addition most medical and pharmacological research is still conducted on men (Epstein 2007).

Furthermore, feminist and decolonial scholars such as Maria Lugones (2007), Oyeronke Oyěwùmí (1997), and Silvia Federici (2014) have each documented how the hierarchical gender system was brought to the 'New World' or the African continent by the colonial conquerors as a habitual way to understand gender and hierarchy where there was none or where the hierarchies were based on different and other categories, like age, experience, or personality. It is important to emphasise in this argument that there is no reason to glorify or romanticise pre-colonial societies as being pure and free from oppression, violence, or hierarchies. This is not the agenda of this chapter. What is important is to counter colonial narratives about gender oppression being a phenomenon from the Global South as if the strict hierarchical ranking order that prevails the current understandings of gender and race is not, in fact, based within a logic that is profoundly reasoned within a modern, scientific Western model of understanding human existence.

Before moving on to the Witch as a liberatory figure and her potentials to decolonise gender, I will elaborate a few important theoretical concepts regarding critical approaches to gender, queer theory and decolonial approaches to gender, and orientalism.

Queer as radical scepticism towards gender

According to scholar and feminist theorist Nina Lykke (2010), queer theory and methodology are central to the development of 'discursive sites of resistance to exclusion, fixity and oppressive meanings of gender/sex' (34). She states:

> In the Anglophone world, 'queer' used to be a popular invective, stigmatizing homosexuals, but queer theory and the queer movements that gained momentum in the 1990s transformed 'queer' into a positive political signifier referring to a broad critical challenge to the dominant heterosexual normativity. (35)

In *Queer Nation. Marginal Sexualities In The Maghreb,* Jarrod Hayes (2000) notes the emphasis on queer as more than a description of sexual acts, that is to say as a verb to signify a critical practice to infiltrate dominant discourses from their political stronghold.

Before proceeding, I want to present my overall approach to the question of gender, sex, and feminist studies, which are subject to great disagreements and conflict even within feminist and gender studies. It is best illustrated by an example from a real life setting: Not long ago, I was invited to give a guest lecture about the subject of witches and cyborgs (and cyborg witches) in a Danish university's gender elective course. In my previous works with witches, I have emphasised how the figure of the witch embodies almost every step of oppression that women have gone through: taking away the wise women's right to be knowledge creators hence rendering women as 'irrational', the attack on women's control over the means of reproduction and population, and the question of domestic work and the subjugation of nature as part of the transition to capitalist modern societies (Ehrenreich and English 1976; Federici 2014; Ghavami Kivi 2019b). But before I had even begun, a student asked why I used the term 'woman' and 'women' if I had a queer perspective to gender. The student's question demonstrates important conflicts within feminist studies: one is the question of gender versus sex, and the other is concerned with the reality of gender, even as we may disagree with it.

The question of gender versus sex is contained in the claim, in some constructionist academic circles, that while sex is a biological fact, gender is a social entity that comes to exist discursively and performatively by a repetitive system of language, signs and demeanours (Butler 1999). That is, gender is largely seen through the lens of Michel Foucault's notion of discourse as "A coherent and strongly bounded area of social knowledge: a system of statements by which the world could be known" (Foucault 1990, 3–35).

The dichotomy of sex as biological and gender as discursive, however, in itself creates the illusion that there is a 'real' material essence in sex, while the discursive approach neglects the bodily realities of gender/sex for actual people of flesh and blood. According to Donna Haraway, the dichotomy is in and by itself a social construction and what we see as 'natural' or 'biological' sex is yet another product of political, cultural, and contextual realities (1991). Lykke describes the problem in more detail:

> Despite the invaluable contributions of gender de/constructionism, it has a problematic side. I have previously described how feminist de/constructionism emerged from a critique of biologically determinist and culturally essentialist gender-conservative discourses. [...] Seen from the perspective of sexual difference theory, gender de/constructionism is problematic because it neglects the bodily irreducibility of sexual difference and the specificity of the female body. Irigaray, for example, emphasizes strongly that to do so is to confirm the hegemonic phallogocentric indifference toward sexual difference. [...] (T)he alternative offered by Irigaray and Cixous is the écriture-feminine/parler-femme perspective and its focus on the relationship between the female body, writing and a disruption of the phallogocentric order.
>
> (Lykke 2010, 107–108)

The 'other', partly included in Lykke's reflections above, allows us to approach the other problem demonstrated by the student's question, namely the one concerned with the reality of gender, even as we may disagree with it: not mentioning the word woman in an attempt to queer gender, would be like not mentioning race as a means to hope that racism will disappear, a phenomenon known as colourblindness (Hübinette and Tigervall 2009). The realities of gender, as well as of racialisation, whether spoken of as such or not, are real in social spheres. Erasing the feminine and women's experience in an attempt to 'queer' can turn out to be a backdoor for making gender oppression invisible and immeasurable as Irigaray and Margaret (2000) and Cixous and Segarra (2010) have pointed out in the quote above. This notion of femininity, that links the non-masculine and the non-white to inferiority and passivity doesn't have to exclusively apply to hegemonic or idealised femininities, like white Western cis-women, but can be extended to include a wide variety of intersections with other categories of oppression and social exclusion. This is explained further in this chapter, through the example of non-hegemonic femininity.

It is this middle way, regarding *both* gender and sex as political and discursive categories, *without* neglecting the embodied and corporal realities of womanhood and femininity in the political landscape of a patriarchal world order, that informs my philosophical approach to the question of gender. This involves questioning and problematising dichotomies such as woman/man, black/white and Western/Oriental, among many others, while at the same time exploring what being marked and read as a feminine body in the world means in accordance to the aims of this chapter to contribute to the abolition of oppressive categories that bind us and limit the freedom of human (and non-human) existence.

Decolonising the witch: how and why?

What is a witch? Is it a person or a profession? Is it an idea? Or is it simply the newest trend among the Global North's watered-down varieties of feminism, where everything can be 'reclaimed' and commercialised as yet another consumer product that signifies transgressive identity and status for the privileged few? What makes the witch such an appealing character? The European witch hunts represent one of the most violently traumatic events in European history and beyond, when read alongside European colonialism. If ever there was a point zero of state-sanctioned terrorism directly targeted at women and other non-conforming people, this was it. So, what or who is the witch from a contemporary perspective and how can we harvest new insights from the current revival and popularity of witches and witchcraft in popular and intellectual culture? These interrogations will lead the way in decolonising the witch.

The witch beyond fear and trendiness

One of the reasons that I have found myself taking issue with the 'trendification' of the witch stems from an encounter with an actual descendent of people who were accused of being witches around 400 years ago. I met her on my recent travels through Europe in search for a specific trail of the witch-hunts that caught my interest some time ago. When I asked her whether she would call herself a witch, she instantly turned pale, lowered her voice, and signalled to me to be cautious of other people within earshot. Around here, she said, people are still terrified of being accused of witchcraft. Under such circumstances, reclaiming is not an option, clearly. Even in our times, before and after the first, second, third, and fourth waves of feminism, witches have been unambiguously associated with evil and everything undesirable in non-conforming womanhood. They were old, sexually unattractive, and vicious—they played games, which made them tricksters—something that lingers on today where women are still seen as hysterical or not to be believed.

This defamation permeates Western culture in various ways, maybe starting with 'Eve in the Garden of Eden' depicted as the original trickster. A consequence of this narrative is exemplified by the #MeToo movement and the need for the hashtag #IbelieveHer, an indicator that women's utterances are rarely accepted and taken for granted (Riva et al. 2011). This disbelief can even have deadly impacts for women and minorities in the health care system. I encountered the roots of this suspicion of women in some of my earlier investigations of the cultural significance of hysteria, a diagnosis given to women who were unmarried and childless in ancient Greece (the cure was claimed to be heterosexual marriage) (Ghavami Kivi 2014). Hysteria, too, has travelled in time. Many women who were diagnosed with hysteria in Sigmund Freud's time would today be undergoing treatment for an autoimmune disease that their contemporaries did not recognise (Delaney, Lupton, and Toth 1977; Scull 2009). Even though hysteria is no longer used as an official diagnosis, I have found that women with hard-to-diagnose physiological illnesses still experience being undertreated and suspected of being 'hysterical', depressed, or simply bored (Ghavami Kivi 2015b). This mistrust towards women leads to grave and often fatal consequences in the health care system.

Witches were accused of tricking young woman into deviation and moral decay—not entirely unlike what feminists have been accused of doing in the 20th century: luring women away from the nuclear family, the institution of marriage, and housework. Witches were not only old and infertile; according to myth, they outright devoured children and knew how to induce abortions with their vast knowledge of medicinal herbs. Consequently, this made them hardcore anti-natalists. Natalism, or the pro-birth position, is believed to have its origins in the era following the Black Death (the Plague) in 14th century Europe, where the population declined dramatically and made the authorities survey births, count population numbers, and gradually push women away from birth work and midwifery as a profession in order to keep track of reproduction and family formation (Federici 2014). It is worth mentioning here that midwives were among various professions under suspicion of witchcraft or sorcery (Ehrenreich and English 1976).

In spite of those groups of women and feminists who have reclaimed witchcraft as a form of political resistance since the 1970s, all these factors play together to maintain the popular depiction of the witch as a gloomy and mainly negative figure. Regardless, there is no denying that the figure of the witch is now (and again) making its way into the mainstream, and being embraced as positive, initially in social media subcultures. The witch has come to symbolise feminist resistance; there are 'Witches Against White Supremacy', 'Witches Against Trump', young woman and queers who embrace being witches as a way of taking pride in outsider identities (Opstrup 2018). There are cyborg witches who reach back to historical knowledge of healing while at the same time hacking modern medicine devices in order to reclaim the medicalised body (Thorburn 2017). And even in academia, witches are slowly clawing their way into the curriculum of gender studies, where previously they have been confined to the fields of history and religion. My own graduate elective course at the Danish Institute for Study Abroad, for example, 'From Witches to Cyborgs: Gender, Race and Resistance', illustrates the connections between colonial history, gendered reproduction, and the current forms of feminist resistance. Witches are here, and they are here for good. That much is beyond doubt.

'The six processes of extermination': witch-hunts as a marker of colonialism

The next question then: How do we go beyond the mainstream properties of the witches' popularity and open new readings that expand our understanding of the witch as a key to our history beyond Europe and beyond a reclaimed feminist icon?

To understand how the witch-hunts go beyond the history of the church versus the people or how the authorities of past times handled 'superstition' and the supernatural, we must take a short history lesson that is absent from most historical accounts of the witch-hunts: their role in the European colonisation of the world. Marxist scholar Silvia Federici has already unveiled much of the history of the witch-hunts and their connection to early primitive accumulation (2017). Adding to that, Julia Suárez-Krabbe has offered a historical context which sees the so-called 'six processes of extermination' as closely connected and indicating the beginning of European expansion, domination, and accumulation of wealth and the starting point of white supremacy. According to Suárez-Krabbe:

> In a period of ten years around the end of 1400, the following key events took place in Europe: the witch hunts had begun; Al-Andalus, the remaining Muslim society had been conquered; the largest expulsion of Jews from Christian soil had been accomplished; colonisation of the Americas had started; the transatlantic slave trade was established; and legal measures to expel Roma populations had been inaugurated. These six historical moments are pivotal to the configuration of coloniality in two ways: on the one hand, they constituted 'whiteness' and its institutions, and by this they also created the realities and loci of enunciation of those they sought to erase.
>
> (2017, 63)

It is important to understand coloniality as more than the occupation of land and enslavement of people for labour. More importantly, coloniality is the creation of a hierarchy based on difference, whether this difference relates to gender, race, culture, religion, or the heterosexist family institutions of modernity. Suárez-Krabbe further elaborates how coloniality goes beyond the matter of geographical occupation of land:

> Coloniality refers to the system of domination that emerged with the European expansion initiated by the Castilian colonial endeavour in the Iberian Peninsula—more specifically the conquest of Al-Andalus and the persecution of the Roma people, the subsequent conquest of the Americas, the witch hunts in Europe and the Americas, and the establishment of the transatlantic slave trade. Coloniality emerged as a local (European) colonial project that spread globally in several interconnected ways. First, it spread through territorial, economic, political and social occupation, theft and control. Second, by the spread and mutual influence of the colonising powers in terms of practices of domination, including their institutionalisation. Third, the global range of coloniality increased by means of racialisation, gender categorisation, sexual domination and labour exploitation and, finally, through cultural, spiritual and epistemic domination. These practices of domination generated a specific articulation between racism, capitalism and patriarchy which is still in play today.
>
> (2017, 59–60)

Other scholars have documented how coloniality and the heterosexist gender hierarchy are closely related, for example, Maria Lugones (2007), documented the existence of a broad pre-colonial gender spectrum among Native Americans where social hierarchies were not centred around one's sex and Oyeronke Oyěwùmí (1997) described the 'ungenderedness' of the pre-colonial Yorúbà world (Coetzee 2018). Silvia Federici (2014) has described how indigenous women in the New World went from pre-Columbian independence to being legally tied to their male partners by the colonisers as a means to bring them with the men to the silver and gold mines, making sure reproductive work—food, cleaning, and mental support for the enslaved men—was done according to the needs of production. These findings point to the notion that the heterosexism and gendered discrimination that were closely linked to the witch-hunts were 'exported' by the colonial logic that was part of European expansion and world domination. In the colonies, accusations of witchcraft were not uncommon. In fact, picturesque descriptions of how indigenous peoples of the New World were 'sorcerers', the word used in witchcraft accusations (Mackay 2009), cannibals, and those who had orgies, were part of the dehumanising propaganda sent back home to Spain and Portugal by the colonisers as a way to justify enslavement, forced labour, and genocide. Witches were not seen as human. They were others.

Sexual others and non-human aliens

Another important strategy of othering the witch by proponents of colonisation and white supremacy uses hypersexualisation to emphasise those accused of witchcraft as non-human others. Witches were accused of and depicted as having sexual relations with the devil and animals, while the broomstick between their legs can be read as a phallic symbol that demonstrates deviation from the prescribed passive feminine sexuality. In this manner, witches share properties with various others, such as people from 'the Orient' (what is now called the Middle East) and Jewish people. The former were depicted as sexually perverted in orientalist accounts of the East, while the latter have been portrayed as greedy and sexually deviant monsters who desired to rape white women, a stigma that was exacerbated in Nazi Germany (Kühne 2018). Using 'the Other' as a mirror image of oneself in order to delineate one's identity is one of the core concepts of Orientalism, developed by Edward Said. In the introduction to his pioneering book, *Orientalism* (1978) he describes the relation of Europe/The West to its nearest neighbour 'The Orient' (Near East or the so-called Middle East of the contemporary) as a means for European self-identification and exploration:

> The Orient is not only adjacent to Europe; it is also the place of Europe's greatest and richest and oldest colonies, the source of its civilizations and languages, its cultural contestant, and one of its deepest and most recurring images of the Other. (1)

Calling 'The Orient', as it is depicted in the European imagination, as invention, Said points out that it is Europe's opposite, aiding it to define itself.

Elaborating on the relationship between Orientalism and sexism, Meyda Yegenoglu (2005) emphasises that this othering has a gendered root that is important to keep in mind when studying orientalism. Imagine the harems, the sexualised ideas of captive women, the stories of rape and sexualised violence that are abundant in stories about 'Muslim women' from a Western point of view. According to Yeğenoğlu (2005) focus should be:

> [...] on the unique articulation of sexual and cultural difference as they are produced and signified in the discourse of Orientalism. I have found that investigations into the question of gender in Orientalism often fall short in recognizing how representations of cultural and sexual difference are constitutive of each other and thus risk reproducing the categorical distinction between the two that feminist theory attempts to combat. With a few exceptions, questions of sexual difference in the discourse of Orientalism are either ignored or, if recognized, understood as an issue which belongs to a different field, namely gender or feminist studies. (1)

In contemporary Europe it is often the sexual 'otherness' of people especially from the so-called Middle East, that is raised by right-wing extremists and increasingly the political centre or social liberals, when they claim that Muslims don't belong in Europe. Evidence of this can be found in the incidence on New Year's Eve in Cologne, where women were attacked, allegedly by refugee men (2016), which was turned into an attack on the nation. A similar discourse is found in the sexual nature of the torture, US soldiers inflicted on Iraqi prisoners in order to humiliate them, while at the same time invading mostly Muslim majority countries in the name of gender equality and improvement of little girls' ability to go to school (Abu-Lughod 2013; Sabsay 2012).

Going back in history, there is also a sexual aspect of anti-Semitism that links the hypersexualisation of 'the Other' in the prosecution of witches, Jewish people and Muslims in both historical and contemporary European politics. Among others, Yvonne Owens (2014) draws a connection between the visual representation of witches and Jewish people in Germany: often with large noses and wicked appearance, they were accused of having 'polluted blood', murdering children, and being guilty of 'heresy'. She also documents the similarities between forced 'cleansing' rituals for women, witches, Jews, and effeminate men, emphasising the sexual nature of these prosecutions in fifteenth and 16th century Germany.

Owens (2014) further documents how the prosecution of witches resembled that of Jews and how processes and rituals of 'cleansing' were similar for Jews, heretics, women, and 'effeminate' men. Typed according to similar figures of 'pollution', Jews and witches were subjected to similar court procedures and suffered comparable 'cleansings', tests, and torture at the hands of the Inquisition. This illuminates the intersection of misogyny, anti-Semitism/orientalism, and sexism that took place in the history of witches. Both witches and Jews were claimed to be carriers of impure blood and governed by Saturn[1]. Not surprisingly, menstruation played a role as well in the argumentation of why women, and people with woman-like properties, like Jews, queers (effeminate men), and others, were impure:

> Throughout the medieval period, authorities expounding on the moral and physical dangers of menstruation had typed feminine blood as an impure substance and stressed that contact with it could have dire consequences. But it is with the increased authority of Aristotelian rationales for the menstrual, moral defect, recovered and assimilated in the thirteenth century, that the 'nature' of femininity could be recodified as a form of treason against the divine order and God's law.
>
> (Owens 2014, 57)

Owens further documents:

> During the late fifteenth and early sixteenth centuries, women— figured as Satan's prime agents—are placed within the frame for sexual maleficia, now reconfigured as witchcraft. Woman's tendency toward transgressive sexuality was thought to be a consequence of her inherently 'defective' nature. (57)

These connections make clear how the history of the witch-hunts cannot be viewed as separate from patriarchal domination in Europe, that took place along racial, gendered, and sexualised lines. In this fashion, witches are accompanied by a large group of people who were marginalised and persecuted as not quite human beings. This is where the queerness of witches in the contemporary comes in.

Queering the witch as a border-crossing migrant

Taking the above into account, my claim is that the witch's return in current times can be seen as an embrace of queer femininity and queer existence. One should be careful with extending the terms 'queer' and 'queerness' to too many unrelated issues, such as national borders or non-LGBTQ+ matters, especially when going back to times before the current concept of queerness existed. However, queerness and queer studies are characterised by a critical approach to power and to the hierarchy that has been established along gender binary lines. Hence, it is an important tool when talking about an entity as gendered as The Witch.

First of all, I see the witch as an anti-woman. In the eyes of the patriarchy, the ideal woman is young, fertile, heterosexual, and willing to take on the task of reproducing the nuclear family and providing (unpaid!) care work for the biological family. The witch is entirely oppositional. The witch is often depicted as old, which brings her past the reproductive stage of most women-identified persons—she is in no way interested in reproducing new citizens for the benefit of any population. She rejects the woman's obligations to (re)produce humans for the nation. She is not the least interested in providing care work for the family nor is she interested in domestic work whatsoever.

One interesting feature about the witch is that she takes one of the most iconic pieces of domestic equipment, the broom, and uses it in a completely different way than intended: to keep the woman in the house and to keep it clean. The witch subverts or queers the broom, so to say. She uses it to get out of the house, to fly, activities that are completely opposite of idealised feminine immobility as seen in fairy tales, like the Sleeping Beauty who lies in a box awaiting rescue by a prince. On the contrary, the witch flies off to distant places, other countries, crossing borders without permission from the border patrol, and goes off to an orgy with her witch friends. The witch, in this way, is also the enemy of the nation state, with its borders and policies of population control. A flying, old, non-heterosexual woman who enjoys herself has no respect for the social contract to work, create families, and strive for personal wealth. By being deliberately anti-nuclear family, anti-nation and anti-border and creating sociality, affinity and kinship on her own terms, socialising with humans and more-than-human however she chooses, the witch undermines each and every one of the rules that are observed by nation states.

Hence, the witch's existence, as I see it, is nomadic; she is a political subject, a fiction that speculates on the destruction of border regimes, of racist and orientalist policies that play women and queers of the Global South as a threat against whiteness and European supremacy. The witch embodies the alienation of many peoples through centuries; people who have been excluded from the notion of humanness. The witch is a figure whom we *must* politicise if we want to advance in our understanding of her. The witch represents multiple intersections of identity and power and embodies different experiences of oppression.

If we want to embrace the witch, if we want to bring her into the light from the darkness, we must remember to embrace the darkness she originates from, and avoid whitewashing, romanticising, or commercialising her history. It is my aim and ambition that—beyond short-lived trends and mainstreaming—the radical, subversive, queer positions of the witch will come to inform our interest in this iconic and very, very real political figure.

★★★

Note: This chapter fur.ther develops ideas presented in an essay published in *Witch Hunt*, a publication accompanying the exhibition Witch-Hunt, edited by Jeppe Ugelvig and Alison Larasyk (Ghavami Kivi, 2020).

Note

1. The Roman god Saturn was said to devour his own children, something that both Jews and Witches have been accused of.

References

Abu-Lughod, Lila. 2013. *Do Muslim Women Need Saving?* Cambridge, MA: Harvard University Press.

Bhattacharya, Tithi. 2017. *Social Reproduction Theory: Remapping Class, Recentering Oppression.* London: Pluto Press.

Braidotti, Rosi. 1994. *Nomadic Subjects: Embodiment and Sexual Difference in Contemporary Feminist Theory.* New York: Colombia University Press.

Butler, Judith. 1999. *Gender Trouble Feminism and the Subversion of Identity.* 10th anniversary edition. New York: Routledge.

Cixous, Hélène and Marta Segarra. 2010. *The Portable Cixous.* New York: Columbia University Press.

Coetzee, Azille. 2018. "Feminism ís African, and Other Implications of Reading Oyèrónké Oyèwùmí as a Relational Thinker." Gender and Women's Studies.

Darwin, Charles. 1859. *On the Origin of Species.* Newburyport: Open Road Media.

Davis, Angela Y. 1981. *Women, Race, and Class.* New York: Random House.

Findlay, Deborah. 1993. "The Good, the Normal and the Healthy: The Social Construction f Medical Knowledge About Women." *Canadian Journal of Sociology* 18, no. 2: 115–135.

Delaney, Janice, Mary Jane Lupton, and Emily Toth. 1977. *The Curse: A Cultural History of Menstruation.* New York: New American Library.

Dobbin, Kristen. 2013. "'Exposing Yourself a Second Time': Visual Repatriation in Scandinavian Sápmi." *Visual Communication Quarterly* 20, no. 3: 128–143.

Ehrenreich, Barbara and Deirdre English. 1976. *Witches, Midwives, and Nurses: A History of Women Healers*. London: Writers and Readers Publ. Co.

Epstein, Steven. 2007. *Inclusion the Politics of Difference in Medical Research*. Chicago: University of Chicago Press.

Eriksson, Elias, Agneta Ekman, Suzanne Sinclair, Karin Sörvik, Christina Ysander, Ulla-Britt Mattson, and Hans Nissbrandt. 2008. "Escitalopram Administered in the Luteal Phase Exerts a Marked and Dose-Dependent Effect in Premenstrual Dysphoric Disorder." *Journal of Clinical Psychopharmacology* 28, no. 2: 195–202.

Federici, Silvia. 2014. *Caliban and the Witch*. 2nd revised edition. New York: Autonomedia.

Federici, Silvia. 2017. "On Primitive Accumulation, Globalization and Reproduction." https://friktionmagasin.dk/on-primitive-accumulation-globalization-and-reproduction-c299e08c3693.

Figert, Anne E. 2005. "Premenstrual Syndrome as Scientific and Cultural Artifact." *Integrative Physiological & Behavioral Science* 40, no 2: 102–113.

Foucault, Michel. 1990. *The History of Sexuality—Vol. 1: An Introduction*. Trans. Robert Hurley. New York: Vintage Books.

Ghavami Kivi, Nazila. 2014. "Mary's bloody knickers." *Friktion*, opsamling 2015–2016. Kbh, friktionmagasin.dk.

Ghavami Kivi, Nazila. 2015a. "De fleste kvinder er ikke normale"—en kritisk analyse af menstruations-gener, Kvinfos webmagasin.

Ghavami Kivi, Nazila. 2015b. "Er smerte hårdest ved mænd?." *Information*, 12.9.2015, https://www.information.dk/moti/2015/09/smerte-haardest-ved-maend.

Ghavami Kivi, Nazila. 2019b. "A Cyborg Is a Witch is a Cyborg is a Witch." *CSPA Quarterly*. No. 24, Winter/Spring 2019, Issue Takeover: Lab for Aesthetics and Ecology, Center for Sustainable Practice in the Arts. https://www.jstor.org/stable/e26629582.

Ghavami Kivi, Nazila. 2019a. "Strangest Things. Mythology, Transformation and Queering Death in Shahrnush Parsipur's Women without Men." In *Unexpected Encounters: Possible Futures*, edited by Kasper Opstrup and Nathaniel Budzinsk. Aarhus: Antipyrine.

Ghavami Kivi, Nazila. 2020. "Queering the Witch as a Nomadic Subject." In *Witch Hunt*, edited by Jeppe Ugelvig and Alice Karasyk. 1st edition Copenhagen: Charlottenborg.

Haraway, Donna. 1990. *Simians, Cyborgs, and Women: The Reinvention of Nature*. London: Routledge.

Hayes, Jarrod. 2000. *Queer Nations: Marginal Sexualities in the Maghreb*. Chicago: University of Chicago Press.

Hübinette, Tobias and Carina Tigervall. 2009. "To Be Non-White in a Colour-Blind Society: Conversations with Adoptees and Adoptive Parents in Sweden on Everyday Racism." *Journal of Intercultural Studies* 30, no. 4: 335–353.

Irigaray, Luce, and Whitford Margaret. 2000. *The Irigaray Reader*. Reprint. Oxford: Basil Blackwell.

Jackson, Zakkiya Iman. 2020. *Becoming Human: Matter and Meaning in an Antiblack World*. New York: New York University Press.

Jensen, Lars. 2018. *Postcolonial Denmark: Nation Narration in a Crisis Ridden Europe*. Milton: Routledge.

Kaminsky, Leah. 2018. "The Case for Renaming Women's Body Parts BBC Future." https://www.bbc.com/future/article/20180531-how-womens-body-parts-have-been-named-after-men.

Karasyk, Alison and Jeppe Ugelvig (eds.). 2020a. *Witch Hunt*. Exhibition at Charlottenborg Kunsthal.

Karasyk, Alison and Jeppe Ugelvig (eds.). 2020b. *Witch Hunt*. Publication at Charlottenborg Kunsthal. Billedeskolernes Forlag.

Kühne, Thomas. 2018. "Introduction: Masculinity and the Third Reich." *Central European History* 51, no. 3: 354–366.

Londa Schiebinger. 1993. "Why Mammals Are Called Mammals: Gender Politics in Eighteenth-Century Natural History." *The American Historical Review* 98, no. 2: 382–411.

Lugones, Maria. 2007. "Heterosexualism and the Colonial/Modern Gender System." *Hypatia* 22, no. 1: 186–209.

Lykke, Nina. 2010. *Feminist Studies: A Guide to Intersectional Theory, Methodology and Writing*. London: Routledge.

Mackay, Christopher S. 2009. *The Hammer of Witches a Complete Translation of the Malleus Maleficarum*. Cambridge: Cambridge University Press.

Martin, Emily. 1996. *The Woman in the Body: a Cultural Analysis of Reproduction*. Milton Keynes: Open Univ. Press.

Mortensen, Kristine Køhler and Marie Maegaard. 2019. "Meeting the Greenlandic people—Mediated Intersections of Colonial Power, Race and Sexuality." *Discourse, Context & Media* 12, 32: 100348.

Nonbo Andersen, Astrid. 2013. "'We Have Reconquered the Islands': Figurations in Public Memories of Slavery and Colonialism in Denmark 1948-2012." *International Journal of Politics, Culture, and Society* 26, no. 1: 57–76.

Opstrup, Kasper. 2018. "Den Utæmmelige Kraft." *K & K* 46, no. 126: 147–176.

Oudshoorn, Nelly. 1994. *Beyond the Natural Body: An Archaeology of Sex Hormones*. London: Routledge.

Owens, Yvonne. 2014. "The Saturnine History of Jews and Witches". *Preternature* 3, no. 1: 56–84.

Oyěwùmí, Oyèrónké. 1997. *The Invention of Women: Making an African Sense of Western Gender Discourses*. Minneapolis: University of Minnesota Press.

Puar, Jasbir K. 2017. *Terrorist Assemblages: Homonationalism in Queer Times*. Durham: Duke University Press.

Riva, Paolo, Simona Sacchi, Lorenzo Montali, and Alessandra Frigerio. 2011. "Gender Effects in Pain Detection: Speed and Accuracy in Decoding Female and Male Pain Expressions." *European Journal of Pain* 15, no. 9: 985.e1–985.e11.

Roberts, Celia. 2002. "A Matter of Embodied Fact: Sex Hormones and the History of Bodies." *Feminist theory* 3, no. 1: 7–26.

Rosaldo, Michelle Zimbalist and Louise Lamphere. 1985. *Woman, Culture, and Society*. Stanford, Calif: Stanford University Press.

Sabsay, Leticia. 2012. "The Emergence of the Other Sexual Citizen: Orientalism and the Modernisation of Sexuality." *Citizenship Studies* 16, no. 5–6: 605–623.

Sabsay, Leticia. 2016. *The Political Imaginary of Sexual Freedom Subjectivity and Power in the New Sexual Democratic Turn*. London: Palgrave Macmillan UK.

Said, Edward W. 1987. *Orientalism*. Harmondsworth: Penguin.

Schulz, Amy J. and Leith Mullings. 2006. *Gender, Race, Class, and Health: Intersectional Approaches*. San Francisco, CA: Jossey-Bass.

Scull, Andrew T. 2009. *Hysteria the Biography*. Oxford: Oxford University Press.

Soble Alan G. 2003. "The History of Sexual Anatomy And Self-Referential Philosophy of Science." *Metaphilosophy* 34, no. 3: 229–249.

Stoler, Ann Laura. 2011. "Colonial Aphasia: Race and Disabled Histories in France." *Public culture* 23, no. 1: 121–156.

Suárez-Krabbe, Julia. 2017. "The Conditions That Make a Difference: Decolonial Historical Realism and the Decolonisation of Knowledge and Education." In *Knowledge and Change in African Universities*, edited by Michael Cross and Amasa Ndofirepi. Rotterdam: Sense Publishers.

Taylor, Diana. 2006. "From 'It's All in Your Head' to 'Taking Back the Month': Premenstrual Syndrome (PMS) Research and the Contributions of the Society for Menstrual Cycle Research." *Sex Roles* 54, no. 5: 377–391.

Thorburn, Elise D. 2017. "Cyborg Witches: Class Composition and Social Reproduction in the GynePunk Collective." *Feminist Media Studies* 17, no. 2: 153–167.

Ussher, Jane M. 2006. *Managing the Monstrous Feminine: Regulating the Reproductive Body.* New York: Routledge.

Valencia, Sayak. 2018. *Gore Capitalism.* South Pasadena, CA: Semiotexte.

von Eggers, Nicolai. 2016. "When the People Assemble, the Laws Go Silent—Radical Democracy and the French Revolution." *Constellations (Oxford, England)* 23, no. 2: 255–268.

Wall, Lewis L. 2006. "The Medical Ethics of Dr J Marion Sims: A Fresh Look at the Historical Record." *Journal of Medical Ethics* 32, no. 6: 346–350.

Yeğenoğlu, Meyda. 2005. *Colonial Fantasies: Towards a Feminist Reading of Orientalism,* Cambridge, England: Cambridge University Press.

6 Abyssal lines in borders, race, and knowledge

A decolonial perspective on the EU-Turkey joint action plan

Avin Mesbah and Sergejs Asilgarajevs

Introduction

In 2015, a significant rise in refugees and migrants arriving on European territory from one year to another consolidated the foundation for what is commonly referred to as the 'European Refugee Crisis' (International Organization for Migration 2016). The situation intensified the state of exception (Agamben 1998) at European Union (EU) border sites, in which constitutional law can be legitimately suspended and progressively replaced by more violent measures. Justified by European anxieties concerning the increased presence of refugees and undocumented migrants, whom Boaventura de Sousa Santos refers to as 'colonial subjects' (2007); physical borders within the EU were erected and accompanied by more aggressive border patrols, while external EU borders were outsourced to fortify the management of 'Fortress Europe' (Frich et al. 2018; Jensen et al. 2017; Jensen and Suárez-Krabbe 2018; Sørensen 2015; Thobo-Carlsen and Tamer 2020). Facing such extraordinary migratory inflows, the EU called upon immediate action to stop 'irregular' and 'illicit' migration across EU borders and the Mediterranean Sea. On October 15, 2015, the EU thus entered a collaboration with Turkey, formally known as the *EU-Turkey Joint Action Plan* (EUTJAP). The EUTJAP was among the first and most comprehensive bilateral policies addressing the 'European Refugee Crisis', with a series of collaborative actions at its centre, supposedly targeting the 'root problems' of the 'crisis' (European Commission 2015).

In contrast to the policy's declared aim of securing the human rights and dignity of Syrian refugees, this chapter argues that the actual 'crisis' addressed in the EUTJAP is the augmented presence of refugees on EU territory, and so it follows that the solution to this crisis is their absence or removal from EU-rope by assigning them temporarily to Turkey, an aspiring EU member-state. As W.E.B DuBois noted in the context of 1920s United States, racism turns people into problems, and along this line, we find it imperative to understand the 'European Refugee Crisis' as a racial crisis, where racialised bodies suffer lethal consequences from the policies to which they are subject. Not only in a social sense but through the logic of epistemic

DOI: 10.4324/9781003172413-7

and legal appropriation and violence, as the policy is informed only by a western-centric knowledge foundation, while the refugee experience and agency are entirely absent and without a voice. Despite the manifestation of modernity/coloniality in the EUTJAP, 'The brute racial fact of this deadly European border regime is seldom acknowledged because it immediately confronts us with the cruel (post)coloniality of the "new" Europe' (de Genova 2018, 1766). The discourse surrounding the 'European Refugee Crisis', and the policies formulated within, is thus de-racialised to such an extent that extremely racialising solutions to what is a *global humanitarian crisis*, are successfully presented as benevolent, compassionate, and merciful, rather than discriminating, exclusionary, and essentially dehumanising. We argue that modern Western thinking, that is abyssal thinking (Santos 2007), is foundational for understanding *whose* problems the policy addresses, and what problems it *disregards* in that very same framing.

In an effort to counter the hegemonic discourse surrounding the 'European Refugee Crisis', this chapter, therefore, undertakes a critical analysis of the EUTJAP from a decolonial perspective and draws attention to the significance of increasingly nuanced understandings of how policies that address the 'European Refugee Crisis' are constructed from within modern Western thinking. The decolonial political discourse analysis applied here thus focuses on the ways policies in contemporary contexts are continuously imbricated in the logic of coloniality/modernity, which inevitably impacts the identification of problems and their solutions (Ahmed 2020, Passada 2019). By applying Boaventura de Sousa Santos' theory of Abyssal Thinking (2007) to the understanding of the EUTJAP, this chapter identifies how the policy follows a colonial tradition of classifying and hierarchising human beings through modern Western knowledge and modern Western law. The policy manages to secure the border interests of the EU, and meet Turkey's desire for a closer collaboration with the EU while reducing the life of a refugee to a transaction in the negotiation of power. Therefore, we argue that the 'EU-ropean' border regime has (re)turned its borders into sites of racialising and dehumanising practices of inclusion and exclusion, turning the refugee body into a *problem* that needs a *solution* rather than a valuable life that needs protection, and a perspective from which the world can be known, and policies developed.

In doing so, we raise questions regarding how the mechanisms of abyssal thinking unfold at EU borders, which are in a state of emergency today. We also raise questions regarding the structural production of nonexistence and sub-humanity through epistemic racism, which is intimately linked to processes of racialisation. Ultimately, this chapter sheds light upon how the concepts of borders, race, and knowledge production are interlinked, and how their meaning and impact on lives effectively reproduce or reinforce racist structures. As such, this chapter is above all concerned with contributing to the decolonisation of the European border regime by elucidating the problems that people fleeing from war, such as the Syrian refugees, face at the borders of Europe.

Colonial/racial lines in knowledge and law

Santos (2007) describes Modern Western thinking as an abyssal thinking, defined by a system of visible and invisible distinctions, which divide social reality into two realms. On this *side of the line*, the abyssal line upholds the metropolitan zone, where the logic of social regulation and social emancipation forms the basis of all modern conflict and development, and enables social repositioning and a negotiation of power (Santos 2007, 2). As we will argue, this logic is foundational in the relationship between the EU and Turkey in the formulation of the EUTJAP and their management of borders. While this logic is fundamentally unthinkable without the '(…) distinction between the law of persons and the law of things', the logic of violence and appropriation prevailing on *the other side of the line*, in the colonial zone, only recognises the law of things—'both human and non-human things' (*ibid.*, 9). Unlike the subjects on *this side of the line* who are recognised in their humanity and agency, colonial subjects are through knowledge and law subjected to racialising processes of exclusion, as is the case for Syrian refugees who are subjected to the EUTJAP (*ibid.*).

By virtue of the abyssal line, the visible distinctions on *this side of the line* are thus separated from—and most importantly founded upon—the invisible distinctions on the *other side of the line* (Santos 2007). Having effectively managed to eliminate the realities of the other side of the abyssal line, modern Western knowledge and modern Western law thus constitute the most accomplished manifestations of abyssal thinking today (*ibid.*). Within modern law, abyssal thinking is reflected in that the legal and illegal states are the only two relevant ways of existing before the law on *this side of the line*, and although assuming universality, this dichotomy leaves out the entire social realm below the abyssal line. Consequently, below the line and in the abyss, we find the lawless, the a-legal, the non-legal, and even the legal or illegal according to non-officially recognised law. Within modern knowledge, this side of the line establishes visible distinctions between everything we consider to be true/false, valid/invalid, human/non-human, and relevant/irrelevant, while there is no factual knowledge on the other side, only beliefs, opinions, or subjective understandings (*ibid.*).

Historically, the global abyssal lines coincided territorially with the cartography of the amity lines that separated the so-called Old World from the New World. These borders defined and supported a global racial hierarchy formulated from a Eurocentric perspective, justifying a violent and systematic dehumanisation of colonial subjects, the appropriation of their land and the de-legitimisation of their epistemology and ontology (Mignolo and Tlostanova 2006). Simultaneously, those situated above the line, the colonisers, and their heirs, were socially recognised for their modernity, superiority, humanity, access to human and legal rights and their ability to produce knowledge (Hansen and Suárez-Krabbe 2018). Although the then visible colonial lines have dissolved, we, like Santos, argue that abyssal lines are *still* constitutive of Western-based political and cultural relations, and interactions in the modern world system, and thus continue to separate the 'coloniser' from the 'colonial' (Santos 2007, 13). The modernity/coloniality of the EUTJAP exemplifies how invisible abyssal lines continue to enable racialising processes in the formulation of policies addressing the 'European Refugee Crisis'. However, these mechanisms are difficult, yet not impossible, to identify as modern Western thinking upholds a system of epistemic racism, which provides the coloniser with the monopoly on defining knowledge and truth, while at the same time working towards silencing and invisibilising countering perspectives (Grosfoguel 2010).

Producing policy from within modern Western thinking

The EUTJAP reflects the dichotomies of modern Western thinking and modern Western law, in that the only two relevant ways of existing before the policy are legal and illegal/illicit, and in the sense that only the EU and Turkey are included in the production of the policy, while the refugees are subject to it and silenced. With regard to the relationship between the EU and Turkey, by recognising Turkey as a part of the international community, the EU situates Turkey within the visible zone on *this side of the abyssal line* (Asilgarajevs and Mesbah 2020). It is also clear that the EU is setting the terms and conditions for this '*joint*' response to the '*crisis*', hence being positioned in the zone of *social regulation*. Turkey is, in turn, contained within the zone of *social emancipation*, through the EU defining Turkey as a '[n]egotiating candidate country' (European Commission 2015, 1). Turkey is being considered a country that could become part of the EU, but must undergo a regulation process to fully comply with the normative of being an EU member state. Therefore, this cooperation becomes a tool for Turkey to negotiate and essentially *emancipate* from its position outside of the EU, while the policy at the same time works as a tool for the EU to *regulate* Turkey's behaviour and border management strategies. This also indicates that currently Turkey does *not* possess the qualities for entering the zone of *social regulation*, that is, becoming a member of the EU.

It is established that the international community possesses the universal values of '*solidarity*', '*human dignity*', and '*humanitarian assistance*'; and by recognising Turkey's agency in applying these values in the management of the crisis, Turkey forms part of *this side of the line*. However, as previously mentioned, the relationship between the EU and Turkey is based upon the logic of regulation and emancipation. If Turkey wishes to fully be part of the EU, the country must adhere to the regulatory terms set by the EU. Thus, while Turkey is placed on *this side of the abyssal line* along with the EU, they are separated by a clear line between the zone of regulation and the zone of emancipation—a line that Turkey has yet to cross. Although Turkey is situated *on this side of the line,* the approach the EU takes in this cooperation is fundamentally based on a colonial model, through which the EU extends its structure, ideas, and power over Turkish territories, by externalising its border management to and beyond Turkey.

The EU as the social regulator on *this side of the line* is identified as everything which embodies truth, legality, universality, modernism, and the traits which indicate superiority and undoubted power, whereas Turkey, placed on the zone of social emancipation, is everything opposite to that of the zone of social regulation (Santos 2007). The EU portrays itself in this policy as possessing all the true qualities of social regulation, by being the one calling for humanitarian actions towards the crisis, by implying that Turkey needs assistance with this crisis and by establishing extensive points of improvements which Turkey needs to undertake as a candidate country for the EU (European Commission 2015). In the formulation of the EUTJAP, Turkey likewise proves to inhabit the zone of social emancipation by holding on to the policy as a tool for moving from an outsider yet included position and crossing into the EU (Asilgarajevs and Mesbah 2020).

The position of the EU and Turkey on *this side of the line* is crucial for their ability and autonomy to identify the *truths* of the 'European Refugee Crisis' along with its main problems and their corresponding solutions. Nonetheless, producing policy from within modern Western thinking implies the absence of perspectives, experiences, and knowledge from subjects inhabiting the *other side of the line,* the refugees, which arguably has immense implications on the formulation of problems and their solutions.

The EUTJAP: whose problems, whose solutions?

The overall narrative of the EUTJAP first and foremost paints the unprecedented 'humanitarian crisis' which not only the EU but the entire 'international community' is facing (European Commission 2015, 1). Based on this observation, the policy rationalises the benefits of the collaboration for the involved parties, followed by the identification of three points of action that target the 'European Refugee Crisis'. Firstly, by addressing the root causes which have led to the massive influx of Syrian refugees; secondly, by supporting Syrians under temporary protection and their Turkish hosting communities; and thirdly, by strengthening the cooperation between the EU and Turkey to prevent irregular migration flows across the Mediterranean Sea and EU borders (*ibid.*). In this context, Turkey is a geographically strategic partner with whom the EU aims to build a stronger migration management strategy in order to control and prevent human flows from moving towards Europe (*ibid.*). The partnership is motivated by a long-term effort between the EU and Turkey to establish a solid, trustworthy relationship that might *eventually* allow Turkey to form part of the EU (ibid.). The outcomes of the policy are thus without a doubt beneficial for both the EU and Turkey.

While identifying and supposedly addressing the causality of the 'crisis', the EUTJAP thus simultaneously serves to define the relationship between the two parties and their positioning in relation to each other. In setting the terms of the agreement, the EU claims that it is of—international and humanitarian—interest for both parties to implement the policy, and accordingly delegates the responsibility to deal appropriately with the problem at hand between the EU and Turkey. The agreement states: 'Challenges are *common,* and responses need to be *coordinated* [...] The EU and Turkey will address this crisis in a spirit of *burden-sharing*' (European Commission 2015, 1; Italics added). The notion of burden-sharing as expressed in the policy not only neglects the humanity of refugees who are reduced to 'burdens', that is, to valueless *things* without agency but it also actively pursues to keep them in that position. Consequently, as refugees are regarded as things, the EU and Turkey have the power to legitimately coerce and manipulate their bodies following their interests, under the guise of humanitarian actions—a narrative that remains generally un-countered (Asilgarajevs and Mesbah 2020).

Subsequent to defining the problem, the policy goes on to reveal the division of labour in tackling the crisis. The responsibility of practical execution is placed on Turkey, while the EU's objective is '[...] to supplement Turkey's efforts in managing the situation of the massive influx of persons in need of temporary protection' (European Commission 2015, 1). Besides outsourcing and strengthening the EU border, this also allows the EU to place large parts of the efforts on Turkey to manage the mass migration towards the EU and at the same time regulate the nation's management of the migration flow.

With regard to addressing the core of the 'crisis', it would be fair to assume that the actual root causes forcing people to flee their countries of origin and seek protection elsewhere would be included. Nonetheless, the policy addresses mainly the migration flows reaching the EU as *the* crisis while stating that what causes this same crisis is a mere *situation* in Syria. Naturally, this is then exactly what Turkey, and to a lesser extent the EU, will *step up* their cooperation to address; the burden of the Syrian refugees seeking temporary protection on their territories.

When mapping the relationship between the parties involved in the policy—the EU, Turkey and the refugees—two lines are identified, a horizontal abyssal line that in subtle manners upholds and unites the EU and Turkey as the *international community* and separates them from the refugees, the target group of the policy, who are situated in the abyss on the *other side of the line.* Between the EU and Turkey, there is however an additional vertical distinction, which sets the grounds for their dynamic relationship defined by the logic of regulation and emancipation. As Santos notes, within modern Western thinking, there is however no visible distinction on the *other side of the abyssal line,* where actions on behalf of the EU and Turkey operate through the logic of appropriation and violence. Both lines are actively either drawn, moved, or strengthened through the *state of exception,* which allows for policies like the EUTJAP to enter and regulate the purpose and functionality of, for example, national borders while disregarding the consequences this might have for the bodies that are subjected to those changes. Due to the non-dynamic logic prevailing in the abyss, refugees are prevented from being included as part of the solution, and are instead positioned as a problem for the EU and Turkey, and reduced to an object of negotiation in their relationship. In this trio, refugees are thus denied a voice of humanity, as they are instead exposed to the logic of appropriation and violence.

In the EUTJAP, the *real* problem for the EU proves to be the mobility of refugees, which Santos refers to as 'the return of the colonial', who due to factors such as globalisation, increased mobility and forced migration are increasingly susceptible to migrating out of the 'colonial' and into the 'metropolitan' zones (Asilgarajevs and Mesbah 2020, 12). Again, the colonial subject is, for Santos, a metaphor for describing those who experience their life as taking place on the *other side of the line* and attempt to rebel against it (Santos 2007). The refugees appearing at the EU border sites are thus causing a situation, in which the EU feels trapped in a shrinking metropolitan space, and reacts by re-drawing the abyssal line (*ibid.*). Therefore, by regulating Turkey by means of the EUTJAP, the EU externalises and strengthens its outer border by redirecting refugees from being a problem in Europe to being a problem in Turkey. Now Turkey is compelled to take all refugees crossing from Turkey into the Greek islands back to Turkey while all costs of the operation are covered by the EU. Essentially, the core of the border management strategy is to keep the Syrian refugees off EU territory and postpone their entrance into Europe. Thus, the physical border becomes a marker of inclusion/exclusion as well as visibility/invisibility, while the border policy reinforces the epistemic colonial difference between beings and nonbeings, sustaining the global racial hierarchy.

The borders as a marker of race

The events of 2015 clearly illuminated the elevated level of anxiety in the political responses to the 'European Refugee Crisis' across the EU. Within a short period of time, borders became contested sites where the sovereign power produced biopolitical bodies through means of surveillance, control, restrictions, differentiation, and classification. However, in one way or another, borders arguably always work towards categorising and classifying to separate one side from another. Regardless of their expression, borders are in their nature inherent to logics of inside and outside and questions about identity and difference. Consequently, borders are not 'natural, neutral nor static but a historically contingent, politically charged, dynamic phenomena' (Vaughan-Williams 2009, 1). Besides being geographical markers, borders are thus also political, subjective (e.g. cultural) and epistemic, and imply the existence of people, languages, religions, and knowledge on both sides linked through relations established by the coloniality of power (e.g. structured by imperial and colonial differences) (Mignolo and Tlostanova 2006).

Indeed, the complexity of modalities in which the processes of inclusion and exclusion are defined, and the formation of biopolitical subjects by the European border regime, remains ambiguous. In times of crises, the border is a space where whoever is subdued to its rules and regulations are situated within a *state of exception* or the logics of abyssal thinking. Under these exceptional measures during a political, social, and cultural crisis, a change of law in border politics must be understood on political and not juridico-constitutional grounds, '[...] then they find themselves in a paradoxical position of being juridical measures that cannot be understood in legal terms, and the state of exception appears as the legal form of what cannot have the legal form' (Agamben 2005, 1). In the state of exception caused by the 'European Refugee Crisis', a process of suspending the current law which determines the surveillance of state borders takes place to produce the zone of distinction in which some bodies are selected through racialising processes of dehumanisation and subjected to increased surveillance. Thus, borders during a state of emergency like the 'refugee crisis', are in a paradoxical position as the *state of exception*.

Although as a consequence, the 'EU-ropean' border regime has turned the borders of the EU into a macabre death scape of black and brown bodies, the situation is pervaded by a total refusal to confront the question of 'race' (de Genova 2018, 1766). Along the lines of abyssal thinking, this causality is linked to the epistemic monopoly on defining the narrative surrounding the immediate actions taken to maintain control over the borders. Acknowledging the actions of appropriation and violence implemented through the EUTJAP confronts the EU with its colonial past and (post-) colonial presence in which the way of treating 'colonial subjects', that is, refugees and migrants, today is rooted. Consequently, this racial crisis has adopted a form where extremely racialising attitudes and structures are invisibilised or normalised through public discourse and policies targeting the situation. Therefore, we stress the importance of understanding racism as a globalised structure that undoubtedly ties back to European colonialism, but localises in particular ways today (Hansen and Suárez-Krabbe 2018). Racism is defined by Ramón Grosfoguel as 'a global hierarchy of superiority and inferiority along the line of the human that has been politically, culturally and economically produced and reproduced for centuries by the institution of the capitalist/patriarchal western-centric/Christian-centric modern/colonial world-system' (Grosfoguel 2016, 10). Thus, although the concept of race might be difficult to articulate in relation to the 'refugee crisis', we argue that the policies addressing the refugee crisis contribute to a gradual but certain dehumanisation of refugees, which allows for the logic of appropriation and violence to apply to them—in this case, exemplified through the EUTJAP. As Loftsdóttir and Jensen stress, racism takes multiple forms in contemporary societies, increasingly attaching itself to features other than race (Jensen and Loftsdóttir 2012). For exactly that reason, it is of utmost importance to 'take racism seriously' and include race and racialisation as analytical categories and tools to deconstruct silencing mechanisms that prevent possibilities for social and political change (Hansen and Suárez-Krabbe 2018). Nonetheless, the human race has throughout history repeatedly demonstrated a tendency to demarcate themselves and other human beings based on selected differences (Asilgarajevs and Mesbah 2020; Fredrickson and Camarillo 2015). Whether divided by religion, gender, or national borders, such differences have either been seen as essential, innate, and unchangeable characteristics—or—cultural, underdeveloped, and adaptable features (Fredrickson and Camarillo 2015). Common for all these departmentalisations is that they inevitably establish distinctions between 'us' and 'them'. Within the world order of modernity/coloniality, and in terms of Santos' notion of residing on this side of the line contra the other side of the line, the concepts of 'race' and 'borders' also have the ability to produce subjects *beyond* these two human categories. Some individuals are thus invisibilised and kept beyond the pale of humanity, as they are subject to a structural production of inexistence (Fredrickson and Camarillo 2015; Grosfoguel 2016; Santos 2007; Suárez-Krabbe 2016). The border works to support the coloniser's (the hegemonic power in the metropolitan zone) efforts to keep the colonial subject on the other side of the border; in this case through adjusting law which becomes a measure of inaccessibility. It is uncertain if the refugee will be able to enter the country, or if/when entering the country, it will be possible for them to remain there under safe terms and conditions. Thus, the border becomes the representation of the zone of distinction between who can participate in the logic of social regulation and emancipation, and who is subdued to the logic of appropriation and violence.

Through policies like the EUTJAP, but also Italy and Libya's Memorandum of Understanding of 2017, which has a similar purpose (Palm 2017), and through the erection of physical borders along with stricter and more violent border controls, the state of emergency has justified diverting from normative procedures and conventions, like the Schengen Agreement. The different facets of the 'European Refugee Crisis' thus reflect the complexity taking place in these zones of indistinction, where the distinction between the inside/outside logic is produced. The 'crisis' manifests how suspending law and invisibilising certain knowledges and perspectives can create a form of life that is easily manipulated and coerced, and how the border becomes a space where this distinction is articulated and executed. The state of exception validates itself through the interaction between European migration policies (zone of being) and refugees (zone of nonbeing), the latter being defined, produced, and maintained by the first, meaning that a mutual connection, by virtue of the separation, is created. This process on the EU border sites creates the distinction between who is allowed to cross the border and who is legitimately kept outside of it—below the abyssal line or as a minimum in the zone of emancipation. Consequently, borders and border policies operate along the same lines as *race*, in terms of who is voiced and heard epistemically and who is included and excluded socially. However, theorising about the EUTJAP from the border allows for an epistemic shift away from seeing people as problems, to engaging with them as people with a perspective and knowledge that can be learned from—also regarding the problems they face—such as borders themselves (Mignolo and Tlostanova 2006). The 'European Refugee Crisis' illustrates the complex issues, links and significance between race, borders and the production of knowledge, and manifests how these concepts have a historic and present significance in producing policy within European modernity.

The contradiction of the humanitarian narrative/discourse

The border does not only represent the geopolitical markings and lines of separation and containment but also racial, cultural, historical, and ideological distinctions between *beings* and *nonbeings* (Fanon 1963). The increased international possibilities of mobility along with factors such as forced migration, economic migration, and contracted displacement have made individuals across the globe more susceptible to migration (Hannam et al. 2006; Santos 2007). Especially the migration of 'colonial' subjects into the (now contaminated) metropolitan zones, is causing the abyssal line to blur and it is awakening anxiety on *this side of the line* in the EU. In 2015, refugees were moving from conflict-affected areas and into Europe at quantities that seemed hard to digest for the political systems of individual member states as well as for the border regime of the EU as a whole. Due to the above-mentioned factors, this is however a tendency that is difficult for the EU to stop entirely, both at the Mediterranean border entry as well as at Turkish border sites, but as the mass presence of colonial subjects on *this side of the line* composes a threat to the established legal, social and epistemic power structures, the EU immediately prioritises to address the issue and reinforce border efficiency. What is happening in the EU is, in general, swarmed by a discourse of *crisis*, a *state of exception*, a penetrating sentiment of *emergency*—all factors which require immediate action, and with great confidence accumulate enough traction to pass policies that directly affect refugees and migrants. This is essential for understanding the relationship between the *representation of the 'crisis'* and the *production of knowledge* formulated through this policy.

Early in the EUTJAP, the EU declared the absolute importance of universal values of *'solidarity'*, *'human dignity'*, and *'humanitarian assistance'* as part of the joint action to solve the "European Refugee Crisis" (European Commission 2015, 1). The EU's humanitarian imagery frames the overall narrative of the EUTJAP, and carefully sets the premise for the EU as the supporting partner with the right experience and knowledge, and for Turkey as the partner that needs to be more proactive and *'step up, intensify, and strengthen'* their efforts in order to fulfil the commitments of belonging to the international community (*ibid.*). The humanitarian discourse on *this side of the line*, effectuated in the visible distinction between the EU and Turkey, is however *only* possible due to the total negation of the violations of the very same discourse and logic on the invisible *other side of the line*. The humanitarian discourse is highly representative of the logic of *appropriation and violence*, which the EU and Turkey apply to refugees, by defining them in terms that situate refugees below the abyssal line, such as being 'burdens' with no agency, hence no humanity and ability to impact their situation. Precisely because they are intended to be kept in the abyss, the appropriation/violence towards them can appear subtle, inexistent, necessary, or insignificant to those above the line.

However, as the logic of appropriation and violence in general terms is presented as *contrary* to Western modernist ideology, the EU must violate its regulations to protect its borders from entering 'colonial' subjects. Therefore, the EU border regime dismisses refugees as humans by reducing them to problems and burdens, though under the guise of protecting their human rights, while in some ways simultaneously violating the same rights: 'Democracy is destroyed to safeguard democracy, life is eliminated to preserve life' (Santos 2007, 16). This also reflects the historical link to Europe's inherent coloniality of 'race' as a system of hierarchisation and access to human rights, where those situated at the top of the human hierarchy are enjoying their adherent human rights, while everything less than human is inferiorised and placed below the abyssal line and stripped of the same rights (Asilgarajevs and Mesbah 2020; Suárez-Krabbe 2016).

Santos defines three major groups that constitute the category of 'colonial subjects', contemporarily intruding the metropolitan societies: 'the terrorist', 'the undocumented migrant worker', and the 'refugee' (Santos 2007, 13). As each of these subjects carries along with them the abyssal line that represents their radical exclusion and non-existence, both today and historically in the coloniality of Europe, their presence creates a condition in which the coloniser, that is, the EU, feels trapped in a shrinking metropolitan space (*ibid.*). The EU, therefore, reacts by re-drawing the abyssal line so that refugees who manage to cross realms from below to above the abyssal line are de facto intentionally kept in the abyss, where they may continuously be subjected to the logic of appropriation and violence. Within this state of exception, democratic, and human rights are restricted under the guise of safeguarding or even expanding them, *within* the metropolitan zone, rather than merely in the colonial zones (*ibid.*). The EUTJAP and its negotiation of refugees represent exactly this: creating a space of appropriation and violence—within the zone of regulation and emancipation. Nevertheless, the tension of regulation and emancipation between the EU and Turkey continues to coexist with the tension of appropriation and violence towards the refugees in such a way that the universality of the first is not contradicted by the existence of the second. This dynamic allows for the invisibilisation of the dehumanisation of refugees, as we argue is the case in the political reactions to the 'European Refugee Crisis'. While the EUTJAP is one such example, policies across various EU member states, who pride themselves on humanitarianism, represent the same mechanisms as the EUTJAP, for example, in the case of Denmark, where deportation and detention centres have become spaces where the logic of appropriation and violence applies *within* the camps towards the 'colonial' subjects, while co-existing with the logic of social regulation and emancipation in the surrounding society. Although criticised for violating basic human rights (Arce and Suárez-Krabbe 2018), these actions within the camps do not contradict the overall humanitarian narrative of the nation due to the mechanisms of epistemic inequality and racism (Grosfoguel 2016; Santos 2007).

Although enshrined in a humanitarian discourse, the EUTJAP strips Syrian refugees from their basic right to mobility and essentially intends to keep them from moving out of their position in the abyss for as long as possible. At the same time, the EU and Turkey are able to profit for their interests and renegotiate their power positions concerning each other through the logics of regulation and emancipation. This is though at the expense of the *humanity of refugees,* which, embodied through this policy, is reduced to that of *sub-human* or *non-existing* (that is in any relevant way), as they are considered merely an object of transaction between the EU and Turkey—a *thing.* Thus, this is a 'joint' action plan, which unites the EU and Turkey while excluding the refugees.

Perhaps a more specific definition of the EU and Turkey's logic in the state of emergency and assessment of the 'refugee crisis', is the application of what Santos (2007) refers to as *the law of things* and *the law of persons.* At the core of modern Western thinking, the colonial subject represents anything that is 'non-Western', 'subhuman' and is thus stripped of agency, silenced, and reduced to a 'thing'. While the dynamic relationship between the logic of social regulation and social emancipation on *this side of the line* is unimaginable without distinguishing between 'the law of persons and the law of thing', the logic prevailing below the line recognises *only* the law of things (Santos 2007, 9). The logic of appropriation and violence is defined by the complexity of extraction of value either cultural and political, for example, customary law and authority in indirect rule, wars, and unequal treaties, and physical, for example, slavery, forced labour, human trading, etc., (Asilgarajevs and Mesbah 2020; Santos 2007). In the EUTJAP, this separation between persons and things is concretely represented through the 'Visa Liberalisation Dialogue with the fulfilment of visa liberalisation roadmap', which would allow Turkish nationals (persons) free visa-less movement through the EU, and the EU-Turkey Customs Union agreement which supports the trade and the flow of goods (valuable things) between the EU member states and Turkey (European Commission 2015). It also represents the *absence* of the same distinction towards the bodies situated below the abyssal line, which are negotiated along the same lines of economic exchange of billions of euros. As the EUTJAP nowhere suggests that refugees are embraced by any laws securing their mobility, as is the case for persons in EU member states and Turkey, it is not a wild suggestion that refugees are stripped down to a bare logic of the law of things. Therefore, by a way of *social regulation* in the relationship with Turkey, the EU can divert refugees to a place of illegality within the zone of social emancipation.

Once the refugee has entered *this side of the line*, they are identified more concretely as *illegal/irregular* or as an undocumented migrant by the European laws of immigration and asylum. Nonetheless, they transit from being *a-legal*, below the abyssal line, to an *illegal* position above the abyssal line and within the metropolitan zone (the EU and Turkey). However, the lack of visa opportunities for migrants and refugees compels them to first arrive on European territory as 'unauthorised' asylum-seekers, meaning they are structurally criminalised and produced as de facto 'illegal migrants', who only thereafter may petition for asylum (de Genova 2018, 1766). Due to the system placing registered refugees and migrants within gated areas such as refugee camps or detention centres, for example in Turkey, their movements are radically limited and, in many cases, these rigid regulations do not dignify their humanity, but instead dehumanises them. Refugees therefore do not gain any immediate freedom by crossing the border and the abyssal line and entering the zone of social emancipation. This also indicates that the border between the EU and Turkey is redrawn and, more importantly, that the EU can enforce its powers onto Syrian refugees *through* Turkey. However, we imagine that it is more likely than not that most refugees view themselves as neither irregular nor illegal, even less so as a *thing* causing a European 'crisis', albeit as human beings fleeing violent conflicts and seeking protection. This legal classification, therefore, is forced onto the refugee but cannot easily be opposed, as the epistemology of Western knowledge impedes exactly that, through producing southern, subaltern, and colonial voices as irrelevant, non-existent, and non-able to counter the hegemonic discourse as more than mere 'subjectivities' (Grosfoguel 2010, 10; Santos 2007, 10).

The power of the EU thus applies not only to *this side of the line* but also to *the other side of the line*. However, the EU's ways of operating are by no means evident, as its actions towards those on the other side are muted in the abyss of inexistence, the subjects not being able, or allowed, to escape the depth of epistemic, ontologic and physical void. Through a logic of violence and appropriation, Syrian refugees are thus represented as a *problem* for the EU border regime that is caused by a mere *situation* in Syria and possesses a great threat to the established order on *this side of the line*. The EUTJAP must thus be seen as a solution, which in a state of emergency capacitates the dehumanisation of the refugees (non-beings), which in turn allows for more extreme and violent measures to be taken into action. Nonetheless, the EUTJAP is fundamentally upheld and enabled by epistemic racism (Fanon 1963, 67; Grosfoguel 2010; Santos 2007).

Conclusions

Modern Western thinking continues to uphold a system of epistemic racism, which provides the coloniser with the monopoly on defining knowledge and truth, while at the same time working towards silencing and invisibilising countering perspectives (Grosfoguel 2010). To counter these intrinsic structural processes of invisibilisation and oppression, this chapter has brought the absent to the foreground to propose an alternative, nuanced understanding of the production of policies that address the 'European Refugee Crisis'. Through a critical analysis of the EUTJAP of 2015, this chapter has questioned and reflected upon *whose* problems the policy addresses, and which problems it *disregards* in that very same framing. We argue that, although enshrined in the dominant discourse of humanitarian aid, values of human dignity, and the importance of human rights, the EUTJAP reduces refugee lives and their presence to *the* problem, while above all serving the individual and shared interests of the EU and Turkey. Operating along the abyssal line, the policy thus manages to separate the coloniser from the colonial, the human from the sub-human, the EU and Turkey from the Syrian refugees. However, this happens in such a way that the human/humanitarian principles of the EU are not compromised by the inhuman/dehumanising practices used in the treatment of refugees. Similarly, the raciality of the EUTJAP can be difficult to address, as it is also invisibilised by hegemonic principles and practices. But as this chapter has argued, the borders erected and fortified during the 'European Refugee Crisis', function as markers of racial differentiation and dictate inclusion and exclusion along the line of the human. Thus, *borders, race,* and *knowledge* are linked in their historic and present significance, as the current state of emergency at EU borders enables racialising practices which are in turn based on a Eurocentric and colonial tradition of producing knowledge.

The underlying factors that enable the discrepancy between the declared (and uncontested) *objective* and the actual *implications* of political responses to the 'European Refugee Crisis', like the EUTJAP, are difficult, yet not impossible to identify. However, this calls for actively countering the sociology of absences, that is, enabled by the modern/colonial world order, and work towards post-abyssal thinking. From a decolonial perspective, the struggle for global social justice at the borders is therefore intimately linked to the struggle for global cognitive justice (Santos 2007). The *ecology of knowledges* demands the recognition of the plurality of knowledges 'without compromising their autonomy', and encourages the idea of the epistemological diversity of the world (Santos 2007; 27). This means assuming equality between diverse forms of knowledge (discarded as 'subjectivities' and as 'nonsense') throughout the world and, most importantly, abandoning the very mechanisms and processes which validate and maintain the epistemological and ontological global hierarchy (*ibid.*).

The deeply embedded de-racialisation of race in the EU-ropean border regime upholds the invisible distinctions enabled by the structure of abyssal thinking. Therefore, we insist that greater attention must be paid to the meaning and functionality of both borders and race in the development towards the post-abyssal. With *critical border thinking,* Mignolo suggests that we can strategically divorce from the epistemic privilege of European modernity, and work towards producing knowledge that empowers and liberates the epistemically oppressed (Mignolo 2012). Parting theoretically from the *border* in international migration policies in general, and in reflecting upon the problems and solutions specific to the 'European Refugee Crisis', would inevitably bring forward alternative understandings of the situation. With regard to the EUTJAP, we could thus bring forward the knowledges from the South, in this case, the voices of the refugees, that are structurally silenced, in that they are not heard nor included in their agency, which produces an absence of their perspectives and epistemic existence, from where they cannot counter the hegemonic discourse. Critical border thinking would therefore imply an epistemic shift from seeing refugees as *problems* to engaging with them as *people* with a perspective and knowledge—also of the problems that they face.

References

Arce, José and Julia Suárez-Krabbe. 2018. "Racism, Global Apartheid and Disobedient Mobilities: The Politics of Detention and Deportation in Europe and Denmark." *KULT: Racism in Denmark*, 15, 107–127.

Agamben, Giorgio. 2005. *State of Exception*. Chicago, IL: The University of Chicago Press.

Agamben, Giorgio. 1998. *Homo Sacer: Sovereign Power and Bare Life*. CA: Stanford University Press.

Ahmed, Yunana. 2020. "Political Discourse Analysis: A Decolonial Approach." *Critical Discourse Studies* Vol. 18 (1), 139–155. Gombe: Routledge.

Asilgarajevs, Sergejs and Avin Mesbah. 2020. *De-racialization of Race in Border Politics: A Decolonial Study of Borders, Race and Knowledge*. Roskilde: Roskilde University.

Dam, Frederikke Amalie, Morten Frich, Merle Baeré, and Frederik Timm Bentsen. 2018. "Mette Frederiksen vil samle Danmark om en udlændingepolitik, som næppe er realistisk", *Information*.

De Genova, Nicholas. 2018. "The 'Migrant Crisis' as a Racial Crisis: Do Black Lives Matter." *Europe, Ethnic and Racial Studies*, 41, 10.

European Commission. 2015. *EU-Turkey Joint Action Plan*. Brussels: The EU.

Fanon, Frantz. 1963. "On Violence." In *The Wretched of the Earth*, 136, translated by Richard Philcox. New York: Grove Press.

Fredrickson, George M., and Albert M. Camarillo. 2015. *Racism: A Short History*. First Princeton Classics edition. Princeton, NJ: Princeton University Press.

Grosfoguel, Ramón. 2010. "Epistemic Islamophobia and Colonial Social Sciences." *Human Architecture: Journal of the Sociology of Self-Knowledge* VIII, 2, 29–37.

Grosfoguel, Ramon. 2016. "What Is Racism?" *Journal of World-Systems Research* 22, 9–15.

Hannam, K., M. Sheller and J. Urry. 2006. "Mobilities, Immobilities and Moorings." *Mobilities*, Vol. 1 (1), 1–22. London: Routledge.

Hansen, Nanna Kirstine Leets and Julia Suárez-Krabbe. 2018. "Introduction: Taking Racism Seriously." *KULT: Racism in Denmark* 1–10.

International Organization for Migration. 2015. "IOM Counts 3,771 Migrant Fatalities in Mediterranean in 2015." *International Organisation for Migration*. Accessed January 4, 2020: https://www.iom.int/news/iom-counts-3771-migrant-fatalitiesmediterranean

Jensen, L., Julia Suárez-Krabbe, Christian Groes, and Zoran Lee Pecic. 2017. *Postcolonial Europe: Comparative Reflections After the Empires*. London: Rowman & Littlefield International.

Jensen, L. and Kristín Loftsdóttir. 2012. *Whiteness and Postcolonialism in the Nordic Region, Exceptionalism, Migrant Others and National Identities.* Farnham: Ashgate Publishing Limited.

Jensen, Lars, Julia Suárez-Krabbe, Christian Groes, and Zoran Lee Pecic. 2018. "Introduction." In Jensen, Lars, Julia Suárez-Krabbe, Christian Groes, and Zoran Lee Pecic (eds.) Postcolonial Europe: Comparative Reflections after the Empires, 1-13. London: Rowman & Littlefield International Ltd.

Mignolo, Walter D. 2012. *Local Histories/Global Designs.* Princeton: Princeton University Press.

Mignolo, Walter D. and Madina V. Tlostanova. 2006. "Theorizing from the Borders. Shifting to Geo- and Body-Politics of Knowledge." *European Journal of Social Theory.* Vol. 9 (2). 205–221. London, Thousand Oaks, CA and New Delhi: Sage Publications.

Palm, Anja. 2017. "The Italy-Libya Memorandum of Understanding: The baseline of a policy approach aimed at closing all doors to Europe?" *EU Immigration and Asylum Law and Policy.*

Passada, María Noel Míguez. 2019. "Discourse Analysis by a Decolonial Perspective." *Advances in Discourse Analysis.* 1–12. DOI: 10.5772/intechopen.8161

Santos, Boaventura de Sousa. 2007. "Beyond Abyssal Thinking: From Global Lines to Ecologies of Knowledges.". *Review, Center for Social Studies,* Coimbra: University of Coimbra 30, no. 1: 1–35.

Suárez-Krabbe, Julia. 2016. *Towards Decolonial Methodology, Race, Rights and Development Alternatives to human rights and Development from the Global South.* London: Rowan & Littlefield.

Sørensen, Lasse Berg. 2015. "OVERBLIK Sådan tackler EU's Lande Flygtningekrisen." DR Nyheder.

Thobo-Carlsen, Jesper and Murat Tamer. 2020. "Nervekrig mellem EU og Tyrkiet afgør migrantaftales skæbne". *Politiken.*

Vaughan-Williams, Nick. 2009. *Border Politics: The Limits of Sovereign Power.* Edinburgh: Edinburgh University Press.

7   Over our dead bodies

The death project, egoism, and the existential dimensions of decolonisation

Julia Suárez-Krabbe

The problem at hand

Many of my students in Denmark are aware of the fact that who they are and what they enjoy and take pleasure in is only possible through the exploitation and death of human and other-than-human beings, and with this they are also acutely conscious of the need to struggle for a better world. Consequently, some of them engage in the environmental struggle, the anti-racist struggle, the queer struggle, the anti-capitalist struggle, or the feminist struggle, while others engage in different, non-organised acts of solidarity, resistance, and treason to the system. The *problem at hand*, which is the focus of this chapter, is that the students, even when engaged in different struggles, often express uncertainty or doubt with regards to the actual *possibility* of another world. In this regard, they seem to be caught up in the 'One-World World'. In a paper engaging the ontological dimensions of Boaventura de Sousa Santos' (2014) notion of 'epistemologies of the south', Arturo Escobar writes:

> Ontologically speaking, one may say that the crisis is the crisis of a particular world or set of world-making practices, the world that we usually refer to as the dominant form of Euro-modernity (capitalist, rationalist, liberal, secular, patriarchal, white, or what have you). Adopting John Law's (2011) compact formulation, we will refer to this world as the One-World World (OWW), that is, a world allegedly made up of a single World, and that has arrogated for itself the right to be 'the' world, subjecting all other worlds to its own terms or, worse, to non-existence; this is a World where only a world fits. If the crisis is then caused by this OWW, it follows that facing the crisis implies transitions towards its opposite, that is, towards the pluriverse.
>
> (Escobar 2016, 15)

Having read and discussed texts such as Santos' and Escobar's, the alternatives presented, for example, the pluriverse and the ecology of knowledges appear too abstract or out of reach to many of the students, who then

DOI: 10.4324/9781003172413-8

pose the important 'how' question, sometimes even wishing for some sort of recipe. My answer is that there is no such recipe, and there cannot be—at least not of the sorts they imagine. Such recipes, or methods, would be counterproductive, for the whole point is that we need to reinsert ourselves into the world, as Fanon (1963) said, and make the path by walking it together with our fellow human and other-than-human beings.[1]

And then, upon hearing this they can get quite frustrated.

Such a frustration is understandable, as walking together requires not only understanding the epistemological and ontological dimensions of decolonisation, but also its existential dimensions. In this chapter, I address these dimensions, their interconnections and nuances that have become clearer and clearer to me, especially in my engagement with students in class, in supervision sessions and when reading their work. This is the first sense in which the paper is a love letter to them, and I write it out of gratitude. It should be clear that the question that concerns me here is not whether my students or I can decolonise or not—this is, to me, a futile discussion. As Frantz Fanon said:

> This huge task which consists of reintroducing mankind into the world, the whole of mankind, will be carried out with the indispensable help, of the European peoples, who themselves must realize that in the past they have often joined the ranks of our common masters where colonial questions were concerned. To achieve this, the European peoples must first decide to wake up and shake themselves, use their brains, and stop playing the stupid game of Sleeping Beauty.
>
> (Fanon 1963, 106)

Taking seriously this Fanonean call, I think of this chapter as part of my conversations and discussions with students in their diversity of histories, engagement, positionalities, experiences, and embodiments. I see these conversations as ways of walking together, at least to the extent that students with whom I no longer have immediate contact may read this, and future students may find some useful input in these ideas. This is the second sense in which this chapter is a love letter to them and I write it to reciprocate what they have given to me.

In the process of thinking through the decolonisation of those of us who are white, mixed, or simply existentially anchored in the One World World, I have played with two interdependent ideas; decolonial historical realism (Suárez-Krabbe 2016; 2017) and relinking (Suárez-Krabbe 2016; 2020; 2021a, 2022). Originally inspired by the works of Lewis Gordon (1995; 2014a; 2014b) and the Mamos,[2] decolonial historical realism resonates with Silvia Rivera's notion of a theory rooted in experience, that is, theory that does not negate one's own history and genealogy in the understanding of the world, and that understands knowledge as something everyone has (Cacopardo 2018).[3] Decolonial historical realism is a radical rootedness, a way of reinserting ourselves into the world, that involves each one of us as historically structured—and structuring—persons belonging to different social and cultural groups. Such rootedness requires engaging in dialogue and learning from our ancestors, and acting as ancestors-to-come; an existentially situated history (Gordon 1995). A notion I learnt from Mamo Saúl Martínez, relinking, on the other hand, involves our relationships to other-than or more-than-human beings besides our ancestors. As such, relinking reintroduces what we otherwise refer to as 'spirituality' and 'magic' into knowledge co-construction. Its particular importance derives from the discussions regarding how to contribute to 'a world in which many worlds are possible', including the historical-political perspectives and relational thinking that such worlds include (Leyva et al. 2015). Like many other similar notions-practices, decolonial historical realism and relinking can be understood as life projects (Blaser 2019) and essentially as struggles against the death project.

My understanding of the death project follows the lead of the Mamos, and also of the indigenous Nasa people in Colombia. With Organizaciones Indígenas de Colombia (OIC), they describe the death project as follows:

> From our origin we are peoples of life. We were born with all living beings. Our Gods taught us to live together in the territory in order to defend the equilibrium and harmony. We are ancestral and aboriginal peoples. The conquerors brought with them their death project to these lands. They came with the urge to steal the wealth and to exploit us in order to accumulate. The death project is the disease of egoism that turns into hatred, war, lies, propaganda, confusion, corruption and bad governments.
>
> (Organizaciones Indígenas de Colombia 2004, my translation from Spanish[4])

Resonating with Achille Mbembe's (2003) 'necropolitics', the death project is concerned with the power and capacity to dispose of life, that is, to the exercise of coloniality, and implies 'the death ethics of war' (Maldonado-Torres 2008), which underlie the legal systems that legitimate it. The death project is then also a way to name the behaviours that accompany these systems, pointing to the international complicity in its continuation. In this regard, the notion of the death project also resonates with Sayak Valencia's 'gore capitalism'. Valencia uses the cinematographic term 'gore' to emphasise the extreme and devastating violence of capitalism; 'the explicit and unjustified bloodshed... the extremely high percentage of viscera and dismemberments, often mixed with organized crime, gender and the predatory uses of bodies, all this through the most explicit violence as a tool for necroempowerment' (Valencia 2010, 15, my translation[5]). However, OIC's notion of the death project adds a crucial aspect to Mbembe's, Maldonado-Torres's, and Valencia's notions as it includes considerations about the negation of Mother World and of spirituality. Consequently, the death

project involves a set of hegemonic practices that emerge out of the insep-
arability of racism, capitalism, patriarchy, and predatory behaviours against
Mother World.

In lectures, I often emphasise that the very structures that we inhabit, as
students and professors in a Danish university, require of us specific ways
of thinking, acting, and engaging which build upon, and perpetuate, the
oppression of others, and the extractivism of materials and knowledges. The
buildings we inhabit are part of the dead mass that now weighs more than living mass on Earth. It is quite telling that today, there are more constructions
like buildings, roads, airports, and other dead products than there is living
biomass on Earth. Such constructions and products weigh around 1.1 teraton while the living biomass weighs 1 teraton. We humans make just 0.01%
of this single teraton of living biomass (Andersen 2020), and the number is
even smaller if we take into account that only a small fraction of humanity is
responsible for the production, making, and consumption of most of the dead
mass. This small fraction of human beings have made these constructions and
products by systematically killing people and other living beings during the
course of the last 500 years. They are the result of a specific way of interacting with Mother World that pretends s/he is nothing other than 'resource'.
Such 'resources' are important elements in an international division of labour
in which people categorised as less-than-human or subhuman serve as the
labour force that extracts the 'resources' so that the 'proper' humans, the
white(ned) man (and his women) can accumulate (Quijano 2000a; 2000b).
In other words, the reality that we are facing today is the result of the ways in
which we, and here I am thinking of those of us who are westernised/whit-
ened, relate to each other and to other earth beings. These ways of relating to
one another are governed by the different ways and degrees in which we are
inserted in the modern colonial racist capitalist patriarchal world system, or
the coloniality of power (Quijano 2000a; 2000b).

Against the egoistic, mono-cultural, and homogenising machinery of the
death project, life projects understand that the very basis of life is difference
and diversity. Expressed in Dan Wildcat's words, 'the way we have to understand the protection of the biological diversity of the planet is *reconnecting* our
human diversity to that biological and ecological diversity' (Wildcat 2014,
my emphasis). In terms of already existing practices in this direction, Wildcat
mentions permaculture and livability. Thinking concretely of the Danish
context and the possibilities of other-worlding here, I add that Denmark
has the highest number of ecovillages per ihabitant on a world scale with 55
existing and 26 emerging (Aagaard 2016). This high number of ecovillages
attests that many people living in the territory of Denmark are significantly
engaged in the protection of biological and ecological diversity. A study
from 2009 that compared CO_2 emissions per person in three ecovillages in
Denmark with the Danish average shows that emissions are 60% lower in the
ecovillages (Hansen 2009). Doubtless, ecovillages also harbour a significant
social and economic potential (Nissen 2020a; 2020b). As types of collective,

communal, and Earth-wise innovation (Escobar 2018), they cannot be judged solely in neo-liberal, materialist, or market terms.

The state of our Earth in Denmark is quite sad, dominated by straight fields of monoculture, straight rivers, and straight woods. Hence, the care and relationality with Earth that ecovillages nurture is important. However, the central point here is that the ecological cannot be delinked from the other crises; racism, inequality, injustice, war, poverty, famine, etc. In Denmark such problems are often disconnected from the climate crisis and the crisis of meaning, and largely placed outside Denmark's—and Europe's—territorial borders, and on specific people produced as 'external' to these borders (Arce and Suárez-Krabbe 2018; Asilgarajevs and Mesbah 2020; Hansen and Suárez-Krabbe 2018). Based on this logic, Denmark has become a de facto apartheid-state by the adoption of laws that legalise attacks, segregation and deportation of refugees, Muslims, and people otherwise produced as non-belonging (Freedom of Movements Research Collective 2018; Özcan and Bangert 2018; Suárez-Krabbe and Lindberg 2019). In the following, I argue that the disconnection between human diversity and biological/ecological diversity is closely linked to an important existential dimension of coloniality; what Organizaciones Indígenas de Colombia (2004) call *the disease of egoism*.

Time, the death project, and the disease of egoism

The modern/colonial global system is an iterative structure—it changes over time, constantly reproducing itself. This is what makes it so powerful, and what makes it appear insurmountable. The existential dimensions of coloniality are key to understand not only the complexity of the problem at hand, but also the vitality of thinking-doing otherwise. In this context, the connection between time, the death project, and the disease of egoism is important, and can be explored by looking into the ways in which our historically constituted colonial system is enforced through specific ways of relating to time and to land into which our European ancestors were forced by other ancestors. The notion of the One-World World as Escobar (2016) uses it refers to the ontological dimensions of the governing modern/colonial world system. As he highlights, the work of Santos addresses its epistemological dimensions, whose effects encompass what Santos has termed 'the indolence of reason' or simply 'lazy reason' (Santos 2004). An important aspect of lazy reason, that allows me to approach the existential dimensions of coloniality, is time. According to Santos, lazy reason 'does not exert itself by thinking of the future because it believes the future is already known—it conceives of the future as a linear, automatic, and infinitely overcoming the present' (Santos 2004, 160). Lazy reason presumes control of the future by history planning and managing nature. Thus, it plans *into* the future, but it never *interacts in,* nor *lives* it. In this way, it *expands the future* so that it is out of reach, distant and, in that sense, irrelevant.

Lazy reason, in other words, detaches us from time. It prioritises and feeds selves who defend that there is nothing to do, that their choices are immediate and disconnected from our future. As Escobar puts it, 'Although taken as the common sense understanding of "the way things are", the One-World World is the result of *particular practices and historical choices*' (Escobar 2016, 21). I emphasise the latter words in this sentence because practice and choice are important to understand the existential dimensions of coloniality. Our experience of time as an accumulation of fleeting moments, and our experience of our individual lifetime as 'our world' are inserted into the extractivist system where many of our choices revolve around criteria such as wasting, using, saving, having or exploiting time. This experience is not 'human nature', it is historically constituted, and it lives on as the continuity of practices based on this way of living and exploiting fellow human and other- or more-than human beings—including time (cf Kudsk 2020). Such ways of living and exploiting time are important ways in which we enforce the death project; indeed, 'Knowledge' can only become 'power' if it can *enforce* its prescriptions (Federici 2004, 141, emphasis added). The death project, then, speaks to the ways in which we enforce the prescriptions of the coloniality of power. The structural enforcement of the death project happens through notion–practices such as human rights and development (Suárez-Krabbe 2016). Such structural enforcement, however, is impossible without the particular practices and historical choices that each person makes. This is why the existential dimensions of coloniality are so essential. Before delving deeper into the existential, it is important to dwell more on the structural contours of the death project.

As Silvia Federici has pointed out in her seminal work about the witch-hunts in the capitalist system, 'life is subordinated to the production of profit' and this demands 'the accumulation of labor-power', that is, the accumulation of people that work *for* the system's sustenance. Such accumulation of labour-power 'can only be achieved with the maximum violence so that, in Maria Mies' words, violence itself becomes the most productive force' (Federici 2004, 16). Against Foucault, Federici further argues, 'torture and death can be placed at the service of "life" or, better, at the service of the production of labour-power, since the goal of capitalist society is to transform life into the capacity to work as "dead labour"' (*ibid.*). Federici shows how the witch-hunts were pivotal to this transformation, which also involved our very relationship to land. In the transition from feudalism to capitalism in Europe, 'It was not the workers—male or female—who were liberated by land privatisation. What was "liberated" was capital, as the land was "free" to function as a means of accumulation and exploitation, rather than as a means of subsistence' (*ibid.*, 75). In other words, the witch-hunts detached our European ancestors, and with them also us, from the land and from each other. Indeed, privatisation needs the individual, the ego.

Thus, far, I have tried to make clear the connection between time, the death project, and the disease of egoism by emphasising how the prescriptions of our historically constituted colonial system are enforced (the death project) through specific ways of relating to time and to land into which our European ancestors were forced—by other ancestors. In Abby Maxwell's words,

> Witch-hunting signified the near-eradication of a Euro-descendent relationship to place, to the more-than human—a bloodstained rupture that has long served racial capitalism. Further, this erasure adhered to the cultural mythology of 'human/nature'—where humans are separate from and superior to nature.
>
> (Maxwell 2020, 11)

As Nazila Ghavami Kivi (2020, 218) notes, however, 'witches are accompanied by a large group of people who were marginalized and persecuted as not quite human beings', and witch-hunting cannot be understood in its full dimensions without taking into account the six processes of extermination that took place during early modernity. Historically simultaneous to the rise of Europe as global imperial power, in a period of ten years around 1492 the witch hunts had begun, Al-Andalus had been conquered and Muslims and Jews were systematically persecuted, murdered, or expelled by the Catholic fundamentalist Spanish powers, the colonisation of territories today known as the Americas was being initiated, the trade with people from the African continent was established, and some of the first legal measures to expel Roma populations had been inaugurated. I have dealt more extensively with these six processes of extermination elsewhere (Suárez-Krabbe 2016), here it is important to highlight that the witch-hunt was part of a larger 'package' that targeted the physical, epistemic, and spiritual lives of many different people: some women, primarily peasant; Muslims, Jews, Indigenous peoples, African peoples, and Roma peoples. With this in mind, I now move on to focus on the ego in its fixation to space/place and time (Federici 2004) to delineate how the disconnection between human diversity and biological/ecological diversity, mentioned earlier by Wildcat (2014), is linked to the disease of egoism.

In order to arrive at the 'I think, therefore I am', the Cartesian 'ego cogito', the subjectivity of our European ancestors emerged first as the 'I conquer, therefore I am', the 'ego conquiro' (Dussel 1995). The I conquer is perhaps the first subject of the death project, as his existence rests upon the exercise of terror, torture, exploitation, expropriation, and other forms of violence. The 'I conquer' is, indeed, both an enslaving and a gendered ego (Dussel 1995, see also Suárez-Krabbe 2016, 86–89; Lugones 2007) that had steadily evolved out of the six processes of extermination mentioned above. The 'I conquer' is an individual already detached from the land that lives a linear temporality. From his perspective, people who may not even have had a gender, nor an ego, and who were close to time and land, were seen as being *behind* in time. The 'I conquer' is incorporated into the 'I think' through the continued engagement in the death project, and through the refinement of the prescriptions of coloniality (see also Maldonado-Torres 2007). This refinement involves the death of the body (Federici 2004, 141), a process tightly connected to the transition to capitalism, in which the body became 'the first machine developed by capitalism' (*ibid.*, 146). In the process of refining the prescriptions of coloniality 'the body as a receptacle of magical powers that had prevailed in the (European) medieval world' was destroyed (*ibid.*, 141), and in close connection to the 'I conquer', emerges the Cartesian model of subjectivity (Dussel 2008), which is pivotal to our current subjectivity. This Cartesian model removed power from the social, and recentred it in the person (Federici 2004, 149–155). Hence:

> The development of self-management (i.e., self-government, self-development) becomes an essential requirement in a capitalist socio-economic system in which self-ownership is assumed to be the fundamental social relation, and discipline no longer relies purely on external coercion. The social significance of Cartesian philosophy lies in part in the fact that it provides an intellectual justification for it. In this way, Descartes' theory of self-management (…) replaces the unpredictable power of the magician (…) with a power much more profitable—a power for which no soul has been forfeited—generated only through the administration and domination of one's own body and, by extension, the administration and domination of the bodies of other fellow beings.
>
> (Federici 2004, 150)

With this quote we can now again return to the problem at hand, which concerns the intimate links between the epistemological, ontological, and existential dimensions of coloniality and decolonisation. Inasmuch as coloniality produced the colonised as subhuman, less than human, or not-even-human, many people are also relegated outside the field of recognition, interaction, and reciprocity (Dussel 1993). Indeed, this is what racism does: it makes people into objects, and thus does not allow them to be part of the social. Of course, this does not mean that people are in fact objects—they are made into such in specific (structural) relationships. In this way, racism means that groups of people are not taken seriously as *human beings* (Gordon 2014a). An example of this that also directs us to the epistemological dimensions of coloniality: currently, the thinking of Africana thinkers, or Arabic thinkers, Indigenous thinkers and many other knowledges from the Global South are largely non-existent in the Danish education system. The education system includes the university, which is one of society's institutions—that is, part of society's structures. When non-white/Southern Global thinking does not exist in the education system *as thinking that is valid and important to enter into dialogue with*, then there is no relationship; if you choose not to meet it, not to engage it, then it does not 'exist' either. So, these thinking traditions are not even structured as 'other traditions'—they are not there at all. Thus, they are not thought of seriously as thinking, and the people behind the thoughts are not seen as (thinking) people either. In other words, the knowledges are erased and those thinkers are invisibilised. This example pertaining to epistemic racism also illustrates the close links between the epistemic, the existential, and the structural. Lewis Gordon's (1999) elaborations on bad faith (mauvaise foi) is central to understand racism in general, also in its epistemic dimensions. Bad faith demands of us that we choose to believe and defend a lie about ourselves and others instead of facing reality. Bad faith is an escape from reality, which is at the same time a renunciation of responsibility in the co-production of reality—in the end, a denial of sociogenesis (Fanon) or design (Escobar 2018) and a rejection of the role one plays in that. If we pretend that racism does not exist, then no one can blame us for reinforcing or playing along with racist structures. Or with regard to the example of the dominance of white-male, Eurocentric, patriarchal thinking in the university: if we choose to pretend that people from the Global South, the colonised, do not produce knowledge, then we do not have to relate to it as knowledge either. This is indeed a prevailing attitude among most colleagues and peers in Denmark.[6]

In the modern colonial racist capitalist patriarchal world system, the *I am* is a self-enclosed individual who situates herself *ontologically* apart from nature. Mother Earth does not exist as I reduce 'it' into nature, resources, ecosystems, mass, private property, etc., all of which are outside the field of (social) interaction. *Epistemologically*, then, I do not listen, feel, sense nor deliberate *with* other-than or more-than human beings (including time and land); I even less value plants, animals, stones, insects, etc., as relatives (Wildcat 2014). They are objects more than subjects; property, not community; resources, not debt; empirical material, not knowledge. As mentioned earlier, coloniality also relegated many peoples outside the field of recognition, interaction, and reciprocity, and hence, the modern/colonial self also practices epistemic racism. *Existentially*, I am a self who is committed to the prescriptions of coloniality, which among others include the idea 'that the world "must" continue as it is presently conceived' (Gordon 1995, 57) as a One-World World. In Madina Tlostanova's words:

> For better or worse, the Marxist ideal human being has never been successfully engineered. However, the winning neoliberal ideal consumer and an utterly fragmented subject alienated from the world and their own self a consumer who lives in the eternal present and whose sphere of desire is completely colonized—has successfully emerged both in its Northern version and its many Southern varieties of second-hand modernity/coloniality.
>
> (Tlostanova 2017, 3)

Indeed, this specific western ego is 'so deeply immersed that it becomes the source of the deepest psychological, psychic and social psychological problems we see modern man confronted with' (Wildcat 2014).

I understand the *disease of egoism* as the existential commitment to the colonial world system, where we continuously choose sealing ourselves away from sociality and relationality, reducing our existence to being a 'human being that maintains structures that militate against human *being*' (Gordon 1995, 83). The colonial, capitalist, racist, patriarchal world demands our existential commitment by de-linking ourselves from Mother World, and to disassociate from each other. These choices and actions that obey coloniality's fundamentally antisocial and antirelational prescriptions add up to become the death project. The disease of egoism emerges when the self becomes the whole. Importantly, the disease of egoism can become a condition of self-implosion to the extent that it also requires us to commit to the lie that embodied reactions to living in an unhealthy society, such as stress and anxiety, are an issue of self-management. As Lewis Gordon noted in an online lecture, however, the problem is rather that 'healthy persons suffer in an unhealthy society', which is why he, along with many others, also speaks about the importance of radical projects of non-narcissistic love: it's not about you or me, but about *us*. This gives new existential meaning to the adage 'get over yourself' (Gordon 2020). Taking the insights pertaining to the effects of the witch hunt in Europe to the development of the *I conquer-I think ego*, those of us who are of Euro-descendent ancestry might usefully choose to go out of our minds and to get over our dead bodies in this existential sense as well. This, the existential dimension of decolonisation, is at the centre of attention in the following concluding section.

Getting over ourselves, out of our minds, and over our dead bodies

Many years ago, Fanon asserted that 'For the colonized, life can only spring up again out of the rotting corpse of the colonizer' (1963, 93).[7] In other writings, I have explored the possibility of the coloniser, or the white man, choosing to become a rotting cadaver (Suárez-Krabbe 2022; 2020). My focus was on white colleagues and peers who, based on the Covid-19 crisis, the Black Lives Matter struggle and the MeToo wave that hit Denmark during 2020, were beginning to see world-transformative possibilities, yet they were clearly avoiding *reintroducing* themselves into the world (Fanon 1963). Instead, I argued, they were choosing to stay in a permanent state of 'white innocence' in Gloria Wekker's (2016) sense. There is an important existential dimension to the possibility of choosing to become a rotting cadaver which connects to the disease of egoism. Because, of course, if all there is, is 'me', I die and then that's it. Remember that in coloniality the self becomes the whole, including time. This is also why many of us tend to think of our choices from within the confines of our own ego's lifetime. There is an aspect to the Fanonean 'rotting cadaver' that now, taking Federici's death of the body into account, acquires new dimensions. Because then, I have been wrong in those previous writings when stating, that the coloniser, or white people, need to choose to become a rotting cadaver so that life can spring up again from us. What is really at stake appears now to be that we need to stop defending the lie that we are *not* rotting cadavers! With the disease of egoism, our bodies are already diseased. To the extent to which the disease of egoism reduces our bodies to machines, stripping us from sociality and relationality; to the extent that we reduce our bodies to labour-power within a system that values us only in terms of our contribution to the labour market; to such an extent then we are indeed already rotting cadavers. To realise our diseased bodies is to get over ourselves and it indeed is also to go out of our minds; it is to reinsert ourselves into the world as part of its processes of becoming, and engage in 'a decolonial politics that centers the human as a relationship and as a situated becoming' (Taylor-García 2018, 2). The 'magic' that was taken out of our bodies and practices of relating to one another during the witch hunt, as Federici contends, consists in the world-making (Lugones 1987) that can now take place over our dead bodies.

As Jane Anna Gordon (2020) reminds us, what the euro-modern language has called magic, is the knowledge-practices of peoples inferiorised such as witches, shamans, Mamos, healers, Taitas, abuelos, and abuelas. Such knowledge-practices include human beings, but human beings are not at the centre. We are part-of knowing-doing. They do not collapse the whole to the self either. Rather, magic expands relationality. In magic, a degree of inexplicable remains, as does a degree of secrecy, for all that is sacred cannot be known, told, nor reduced to the level of reality in which things can be explained. A lot of what is cosmological is also called magical, and in many communities the wise people, healers or sages are those among them who have accepted and received special guidance to engage in the knowledge-practices that emerge in relationships with other than- and more-than-human beings.[8] This does not make the shaman, or the magician, the one ruler above others; they are, instead, teachers who guide others—to the extent that we choose to engage in such processes of knowing and relating. Neither does this mean that white people do not have the magic. Indeed, according to the Mamos, we all have the ability to relink but, as Saúl Martínez says, we have been taught to use only five of our senses, and to use those in ways framed from within the One World-World. The other senses, however, are still there but, like unused limbs, they are weakened (Suárez-Krabbe 2016, 169). To engage in the knowledge-practices that emerge in relationships with other than- and more-than-human beings, we therefore need to practice, know and sense otherwise. Importantly, then, all children, even those born and raised in the West, have 'the magic'. However, it is disciplined and educated out of us, in the terms I have used in this chapter, we have to greater or lesser extents been socialised into becoming rotting corpses.[9] In a relational world, magic is not something an individual does, magic is that which emerges *as* a relation and *in* relation to other beings. Magic is a relationship in which we live and interact with time otherwise, beyond the linear conception of time in Santos' (2005) lazy reason. Magic, in this sense, reintroduces us in different temporalities which enables us to see the alternatives that are to be found in the horizon of concrete possibilities which we forge through our interactions with other human and other-than-human beings (see also Rivera 2012). Significantly, in this horizon of concrete possibilities we also engage in interaction *with* time (Kudsk 2020). Inspired by abuelo Rodolfo, Kudsk introduces the notion of 'ecology of history'. This notion breaks with lazy reason showing how:

> the social (historytelling) can be understood in new/other ways (in relation to the dominant, western), when time and nature are taken seriously as factors for this. The lived experience is history is tradition. History is that we humans have forgotten to ask for permission, we have forgotten to give back. Ecology of history shows us that this historytelling is just as valid as the one written in the books of historians. This is the decolonisation of history understood as the Western idea of progress through the constant exploitation of the earth's 'resources'[10].
>
> (Kudsk 2020, 28)

Historytelling, as I understand Kudsk, implies to be *existentially* situated in relationship to time and nature; this is what magic can be. Magic places the determinants of social action in the realm of the social and relational, out of reach and control of one single person. Consider, for instance, the power of prophecies:

> Prophecies are not simply the expression of a fatalistic resignation. Historically they have been a means by which the 'poor' have externalized their desires, given legitimacy to their plans, and have spurred to action. Hobbes recognized this when he warned that 'There is nothing that… so well directs men in their deliberations, as the foresight of the sequels of their actions; prophecy being many times the principal cause of the events foretold (Hobbes, *Behemot*, Works VI: 399)'.
>
> (Federici 2004, 143)

Prophecies are ways of interacting with time and of making history foretold, and we know that many peoples around the world engage in this kind of long-term interaction with time, history, and each other as ancestors and future generations. In their recipe for 'Spell Casting', CrimethInk ex-Workers' Collective (2005) explain this magic power of radical sociality in close relationship to time with the following words:

> To pull off a revolution on any scale, you have to be able to believe outside the box. Reality, both present and future, is created by mass consensus […] Even a small group of people who believe against the grain can call an entire world-system into question, not to mention liberate themselves from its supposed inevitabilities. If the alternate world they

consider themselves to inhabit is convincing, and more appealing than the one everyone accepts, the future itself can be hijacked by the desires this minority trusts and thus unleashes. To speak on a smaller scale, perception and reality influence one another, and believing that something is possible is generally a prerequisite for being capable of bringing it about. In this sense, whether the desire for revolution or anything else is 'realistic' is a moot point: for the individual who does not wish to cripple herself, the question is not what to believe in as 'the' truth but what beliefs render which truths possible.

(2005, 501–502)

Spell casting as it is further described is to embody and enact magic: it is a way of conceiving the getting over ourselves, out of our minds, and over our dead bodies, which I have been writing about here. But spell casting without the ecology of history—the existential situatedness in relationship to time *and* nature that Kudsk conceptualises with the teachings of abuelo Rodolfo—will risk to remain limited. Dan Wildcat additionally reminds us that inasmuch as 'we existed as human kind on this planet as tribal nations for thousands of years before we became members of the modern nation states', then it is important that we remember that 'everyone has ancestry much deeper and richly grounded in tribal traditions' than in the current colonial system. In this sense, as abuelo Rodolfo says, we Euro-descendent people have indeed forgotten to 'ask for permission, we have forgotten to give back' (Kudsk 2020, 28). Wildcat's and abuelo Rodolfo's words resonate with Tlostanova's description of decolonial design as a:

creative and dynamic reflection and realization of the people's forgotten and discarded needs, wishes and longings, which would be inevitably linked to the local cosmologies, ethics and systems of knowledge seen not as the dead and museumized past, or as a conservative fundamentalist dystopia, but as a living and breathing present and a promise for the future.

(Tlostanova 2017, 5)

At the beginning of this chapter, I mentioned the Danish ecovillages and their transformative potential. Indeed, according to Ditlev Nissen (2020b), practices such as those described by Tlostanova are already taking place in the ecovillages. Processes, as those the students are engaged with, and the elaborations on magic, the ecology of history, prophecy, and spell casting all point to the pluriverse not as an impossible future, but as a vast diversity of world-making practices. When I emphasise practice, it is because the world we live in now, the dominant colonial system is the result of a specific way of world-making (the death project) that relies on our choices and practices to exist. In the same way as colonisation is a verb—it is what people have done and continue doing—so is decolonisation. So, this chapter is my way of answering the students' question regarding the 'how' of the pluriverse: a world of many worlds is not only possible, it is already taking place (Lugones 1987). This is not to deny the fact that we are living very difficult times, nor to affirm that the death project is gone, nor to pretend that we are all exposed to the premature death produced through the global racist structures in the same way nor to the same extent (Gilmore 2007). On top of already existing crises, the Covid-19 pandemic has thrust many of us into a deep existential crisis, at least to the extent that there is a clear sense of a 'cosmic tectonic shake', an end of the world as we (think) we know it, a generalised sense of 'unsettle-ment' (Tlostanova 2020). This is also true to those of us who have ridden the corona crisis in first class (Drachmann 2020). As different prophecies around the world have known, however, this is the end of a cycle, an historical era, of an epoch, not the end of the world. Ends are also beginnings and in-be-tweens where our worlds are (re-)emerging. As the colonial system continues to fall apart, the struggle consists not only of de-linking from the death pro-ject and struggling against it, but of re-linking to life projects. At this stage, at least as reflected in my students, we are increasingly becoming conscious that our struggles are attempts to co-create a new world in which many worlds fit. So, the question is not whether the pluriverse is possible, the question is 'about commitment, even without knowing the results' (Gordon 2020). Taking the pluriverse seriously, reinserting ourselves in it, is then a necessary task in Denmark too, despite the resistance coming from western dominant thinking and its scientific communities. Students are increasingly taking these insights with them, applying and practicing other-worlding, broadening their reach, and our worlds. A guiding question in this concern, which we can pose ourselves and each other as we walk together, is: what historical moment are we living/making?

Notes

1. Making the path by walking is a common notion in Latin American decolo-nial traditions, my use of it is highly inspired by Vasco's notion of *andar* (2002; 2007). It should be noted that there are many other ways to understand the doing-being-thinking-feeling otherwise towards another world, see for instance the con-tributions in Leyva et al. (2015).
2. Mamo is the denomination used by the indigenous peoples (Kankuamo, Kogi, Arhuaco, and Wiwa) in Sierra Nevada de Santa Marta, Colombia, to name their wise elders.
3. Close paraphrase of the following sentences in Spanish: 'Que tiene raíz en la expe-riencia, que no niega la historia propia ni la genealogía propia para la comprensión del mundo. Y que concibe el saber como algo que porta todo ser humano.'
4. 'Desde nuestro origen somos pueblos de la vida. Nacimos con todos los seres vivos. Nuestros Dioses nos enseñaron a convivir en el territorio para defender el equi-librio y la armonía. Somos pueblos ancestrales y originarios. Los conquistadores trajeron a estas tierras su proyecto de muerte. Vinieron con afán de robarse la riqueza y explotarnos para acumular. El Proyecto de Muerte es la enfermedad del egoísmo que se vuelve odio, guerra, mentiras, propaganda, confusión, corrupción y malos gobiernos.'

5. 'Tomamos el término gore de un género cinematográfico que hace referencia a la violencia extrema y tajante. Entonces, con capitalismo gore nos referimos al derramamiento de sangre explícito e injustificado (como precio a pagar por el Tercer Mundo que se aferra a seguir las lógicas del capitalismo, cada vez más exigentes), al altísimo porcentaje de vísceras y desmembramientos, frecuentemente mezclados con el crimen organizado, el género y los usos predatorios de los cuerpos, todo esto por medio de la violencia más explícita como herramienta de necroempoderamiento.' (Valencia 2010, 15).

6. Denmark's higher education system is a significantly closed one that continuously re-creates and re-validates colonial thinking; this happens from the highest governmental structures through educational policies, to students who do not have access to or qualified assessment to engage Southern knowledges in their academic programmes.

7. The original translated text uses the word 'native' and 'settler' instead of 'colonised' and 'coloniser', which I have inserted instead for ease of understanding, and to avoid the inversion of terms in the case of Euro-descendent people, who are not native in the sense Fanon uses the word.

8. Largely paraphrase of Gordon's talk (2020).

9. I am thankful to Zuleika Sheik for reminding me of this, helping me to clarify the above points. She adds: 'If anything, using the example of the witches of Europe, points to magic as inherent to being/becoming and as such more work needs to be done to reignite the connection (whereas for some others say in the Global South, the flame was never extinguished).' Personal communication, March 18, 2021.

10. ...det sociale (historiefortællingen) kan forstås på nye/andre (i forhold til den dominerende, vestlige) måder, når tid og natur tages seriøst som faktorer herfor. Den levede erfaring er historie, er overlevering. Historien er det, at vi mennesker har glemt at bede om tilladelse, vi har glemt at give tilbage. Historieøkologi viser os, at denne historiefortælling er lige så gyldig som den, der står skrevet i historikernes bøger. Dette er afkolonisering af historien forstået som den vestlige idé om fremskridt gennem stadig udnyttelse af jordens 'ressourcer'.

References

Aagaard, Niels. 2016. "Formandsberetning til generalforsamling for Landsforeningen for Økosamfund." LØS generalforsamling 23. april 2016 i Hvalsø under Folketræffet. http://okosamfund.dk/wp-content/uploads/Formandens-beretning.pdf.

Andersen, Ellen Ø. 2020. "Kloden passerer en milepæl: De døde, menneskeskabte ting vokser og cokser og vejer nu mere end alt Jordens levende liv" *Politiken*, December 11, 2020. https://politiken.dk/viden/Viden/art8027914/D%C3%B8d-menneskeskabt-masse-har-overhalet-Jordens-levende-natur.

Arce, José and Julia Suárez-Krabbe. 2018. "Racism, Global Apartheid and Disobedient Mobilities: The Politics of Detention and Deportation in Europe and Denmark." *KULT. Postkolonial Temaserie*. 107–127.

Asilgarajevs, Sergejs and Avin Mesbah. 2020. *De-racialization of Race in Border Politics: A Decolonial Study of Borders, Race and Knowledge*. Master's thesis in Cultural Encounters and International Development Studies, Roskilde University.

Blaser, Mario. 2019. "Life Projects." In *Pluriverse: A Post-development Dictionary*, edited by Ashish Kothari et al., 234–236. New Delhi: Tulika Books.

Cacopardo, Ana. 2018. "Nada sería posible si la gente no deseara lo imposible". *Entrevista a Silvia Rivera Cusicanqui*. Andamios 15, no. 37: 179–193.

CrimethInk ex-Workers' Collective. 2005. "Spell Casting." In *Recipes for Disaster. An Anarchist Cookbook a Moveable Feast*. Olympia: CrimethInk. Far East.

Drachmann, Hans. 2020. "Danskerne kører gennem coronakrisen på 1. klasse" *Politiken*, December 18, 2020. https://politiken.dk/indland/art8040157/Danskerne-k%C3% B8rer-gennem-coronakrisen-p%C3%A5-1.-klasse.

Dussel, Enrique. 1993. "Eurocentrism and Modernity (Introduction to the Frankfurt Lectures)." *Boundary 2* 20, no. 3: 65–76.

Dussel, Enrique. 1995. *1492. El encubrimiento del Otro. Hacia el origen del mito de la modernidad*. La Paz: Biblioteca Indígena.

Dussel, Enrique. 2008. "Meditaciones anti-cartesianas: sobre el origen del anti-discurso filosófico de la Modernidad." *Tabula Rasa* 9: 153–197.

Escobar, Arturo. 2016. "Thinking-feeling with the Earth: Territorial Struggles and the Ontological Dimension of the Epistemologies of the South." *Revista de Antropología Iberoamericana* 11, no. 1: 11–32. https://10.11156/aibr.110102e.

Escobar, Arturo. 2018. *Designs for the Pluriverse. Radical Interdependence, Autonomy, and the Making of Worlds*. Durham and London: Duke University Press.

Fanon, Frantz. 1963. *The Wretched of the Earth*. Translated by Richard Philcox. New York: Grove Press.

Federici, Silvia. 2004. *Caliban and the Witch*. New York: Autonomedia.

Freedom of Movements Research Collective. 2018. *'Stop Killing Us Slowly'. A Research Report on the Motivation Enhancement Measures and the Criminalization of Rejected Asylum Seekers in Denmark*. Freedom of Movements Research Collective, Copenhagen. http:// refugees.dk/media/1757/stop-killing-us_uk.pdf.

Ghavami Kivi, Nazila. 2020. Queering the Witch as a Nomadic Subject. In *Witch Hunt. A Reader on the Nordic Witchcraft Trials*, edited by Alison Karasyk and Jeppe Ugelvig, 214–220. Copenhagen: Billedkunstskolernes forlag.

Gilmore, Ruth. 2007. *Golden Gulag: Prisons, Surplus, Crisis, and Opposition in Globalizing California*. Berkeley, Los Angeles, London: University of California Press.

Gordon, Jane Anna. 2020. "The Magic Hands are Finally Only the Hands of the People: Continuing to learn with Fanon at 95." Online lecture delivered at the Caribbean Philosophical Association's Fanon at 95 event, July 1-20, 2020. July 14, 2020. Last accessed December 10, 2020. https://www.facebook.com/CaribPhil/videos/ fanon-at-95-july-14th-celebration/728856257848290/.

Gordon, Lewis. 1995. *Fanon and the Crisis of European Man. An Essay on Philosophy and the Human Sciences*. New York and London: Routledge.

Gordon, Lewis. 1999. *Bad Faith and Antiblack Racism*. New York: Humanity Books.

Gordon, Lewis. 2014a. "Black Existence in Philosophy of Culture." *Diogenes* 59, no. 3–4: 96–105.

Gordon, Lewis. 2014b. *What Fanon Said? A Philosophical Introduction to His Life and Thought*. London: Hurst.

Gordon, Lewis. 2020. "What to Do in Our Struggle to Breathe: Fanon's Relevance in Our Time of Multiple Pandemics." Online lecture delivered at the Caribbean Philosophical Association's Fanon at 95 event, July 1–20, 2020. July 13, 2020. Last accessed December 10, 2020. https://www.facebook.com/CaribPhil/videos/fanon-at-95-july-13th-celebration/593647091339732/.

Hansen, Kaj. 2009. "Reduction of CO_2 from 3 different eco-villages in Denmark." *Løsnet* 61–62: 4–6. https://issuu.com/okosamfund/docs/l__snet_61-62__2.del_lille.

Hansen, Nanna Kirstine Leets and Julia Suárez-Krabbe. 2018. "Introduction: Taking Racism Seriously." *KULT: Racism in Denmark* 15: 1–10.

Kudsk, Dea. 2020. *At genordne tænkningen. En studie i forbindelser mellem undertrykkelse af viden og destruktion af verden.* Masters thesis, Cultural Encounters. Department of Communication and Arts, Roskilde University.

Leyva, Xoxitl et al. 2015. *Prácticas otras de conocimiento(s). Entre crisis, entre guerras.* Tomo I, II, III. San Cristóbal de Las Casas: Cooperativa Editorial Retos.

Lugones, María. 1987. "Playfulness, 'World'-Travelling, and Loving Perception." *Hypatia* 2, no. 2: 3–19.

Lugones, Maria. 2007. "Heterosexualism and the Colonial/Modern Gender System." *Hypatia* 22, no. 1: 186–209.

Maldonado-Torres, Nelson. 2007. "On the Coloniality of Being: Contributions to the Development of a Concept." *Cultural Studies* 21, no. 2–3: 240–270.

Maldonado-Torres, Nelson. 2008. *Against War: Views from the Underside of Modernity.* Durham, NC and London: Duke University Press.

Mbembe, Achile. 2003. "Necropolitics." *Public Culture* 15, no. 1: 11–40.

Maxwell, Abby. 2020. "On Witches, Shrooms, and Sourdough: A Critical Reimagining of the White Settler Relationship to Land." *Journal of International Women's Studies* 21, no. 7, 8–22. https://vc.bridgew.edu/jiws/vol21/iss7/2.

Nissen, Ditlev. 2020a. "Økosamfund som bæredygtige fremtidslaboratorier." *Levende Lokalsamfund.* January 19, 2020. https://levendelokalsamfund.dk/oekosamfund-baeredygtige-fremtidslaboratorier/.

Nissen, Ditlev. 2020b. "Økosamfund som læringsrum og forskningsfelt." *Levende Lokalsamfund* February 18, 2020. https://levendelokalsamfund.dk/oekosamfund-laeringsrum-forskningsobjekt/.

Organizaciones Indígenas de Colombia. 2004. "Propuesta política y de acción de los pueblos indígenas. Minga por la vida, la justicia, la alegría, la autonomía y la libertad y movilización contra el proyecto de muerte y por un plan de vida de los pueblos." Last accessed December 10, 2020. http://anterior.nasaacin.org/index.php/2010/06/04/propuesta-politica-de-los-pueblos/.

Özcan, Sibel and Zeynep Bangert. 2018. "Islamophobia in Denmark: National Report" In *European Islamophobia Report 2018*, edited by Enes Bayraklı and Farid Hafez. Istanbul, SETA, 251–282: http://www.islamophobiaeurope.com/wp-content/uploads/2019/09/DENMARK.pdf.

Quijano, Aníbal. 2000a. "Coloniality of Power, Ethnocentrism, and Latin America." *Nepantla, Views from the South* 1, no. 3: 533–580.

Quijano, Aníbal. 2000b. "Colonialidad del Poder y Clasificación Social." *Journal of World-Systems Research* 4, no. 2: 342–386.

Rivera, Silvia. 2012. Ch'ixinakax utxiwa: A Reflection on the Practices and Discourses of Decolonization. *South Atlantic Quarterly* 111, no. 1: 95–109.

Santos, Boaventura de Sousa. 2004. "A Critique of Lazy Reason: Against the Waste of Experience." In *The Modern World-System in the Longue Durée*, edited by Immanuel Wallerstein, 157–97. London: Paradigm Publishers.

Santos, Boaventura de Sousa. 2005. *El milenio huérfano. Ensayos para una nueva cultura política.* Madrid: Trotta.

Santos, Boaventura de Sousa. 2014. *Epistemologies of the South. Justice against Epistemicide.* London and New York: Routledge.

Suárez-Krabbe, Julia. 2016. *Race, Rights and Rebels: Alternatives to Human Rights and Development from the Global South.* London: Rowman & Littlefield International.

Suárez-Krabbe, Julia. 2017. "The Conditions that Make a Difference. Decolonial Historical Realism and the Decolonisation of Knowledge and Education." In *Knowledge and Change in the African Universities*, edited by Michael Cross and Amasa Ndofirepi, 59–80. London: Sense Publishers.

Suárez-Krabbe, Julia. 2020. "Relinking as healing: Ruminations on crises and the radical transformation of an antisocial and antirelational world." Last accessed December 10, 2020. https://www.convivialthinking.org/index.php/2020/10/05/relinking-as-healing/.

Suárez-Krabbe, Julia. 2021a (forthcoming). "Relinking as a Knowledge-Practice. Conversations towards the Decolonization of Knowledge and Being." In *Decolonial Reconstellations: Critical Global Studies Beyond Eurocentrism*, co-edited by Laura Doyle and Mwangi Wa Githinji. Under review at Duke University Press.

Suárez-Krabbe, Julia. 2022 (forthcoming). "Racism, Sociogeny and (Im)possible Decolonization. Reflections on the Crisis of European Man in Denmark." In *Fanon and the Crisis of European Man* (anniversary edition), edited by Lewis Gordon. To be published with Routledge.

Suárez-Krabbe, Julia and Annika Lindberg. 2019. "Enforcing Apartheid? The Politics of "Intolerability" in the Danish Migration and Integration Regimes." *Migration and Society. Advances in Research.* 2, no. 1: 90–97.

Taylor-García, Daphne. 2018. *The Existence of the Mixed Race Damnés. Decolonialism, Class, Gender, Race.* London: Rowman and Littlefield International.

Tlostanova, Madina. 2017. "On decolonizing design." *Design Philosophy Papers* 15, no. 1: 51–61. DOI: 10.1080/14487136.2017.1301017.

Tlostanova, Madina. 2020. Of Birds and Trees: Rethinking Decoloniality Through Unsettlement as a Pluriversal Human Condition. *Echo. Rivista interdisciplinare di comunicazione* 2: 16–27.

Valencia, Sayak. 2010. *Capitalismo Gore.* Spain: Melusina.

Vasco, Luis Guillermo. 2002. *Entre selva y páramo. Viviendo y pensando la lucha india.* Bogotá: Instituto Colombiano de Antropología e Historia.

Vasco, Luis Guillermo. 2007. "Así es mi método en etnografía." *Tabula Rasa* 6: 19–52.

Wekker, Gloria. 2016. *White Innocence. Paradoxes of Colonialism and Race.* Duke University Press.

Wildcat, Dan. 2014. "On Culture and Community." Bioneers Indigenous Knowledge. Youtube. April 14, 2014. Last accessed December 10, 2020. https://youtu.be/AwsN8aDjQrU.

Index

Note: Italicized page numbers refer to figures. Page numbers followed by "n" refer to notes.

Printed in the United States
by Baker & Taylor Publisher Services